# Lecture Notes in Physics

Edited by J. Ehlers, München, K. Hepp, Zürich
R. Kippenhahn, München, H. A. Weidenmüller, Heidelberg
and J. Zittartz, Köln
Managing Editor: W. Beiglböck, Heidelberg

98



# Nonlinear Problems in Theoretical Physics

Proceedings of the IX G.I.F.T. International Seminar on Theoretical Physics, Held at Jaca, Huesca (Spain), June 1978

Edited by A. F. Rañada

Springer-Verlag
Berlin Heidelberg New York 1979

**Editor**

Antonio F. Rañada
Facultad de Ciencias Fisicas
Universidad Complutense de Madrid
Madrid
Spain



ISBN 3-540-09246-3 Springer-Verlag Berlin Heidelberg New York
ISBN 0-387-09246-3 Springer-Verlag New York Heidelberg Berlin

Library of Congress Cataloging in Publication Data. International Seminar on Theoretical Physics, 9th, Jaca, Spain, 1978. Nonlinear problems in theoretical physics. (Lecture notes in physics ; 98) Bibliography: p. Includes indes. 1. Nonlinear theories--Congresses. 2. Mathematical physics--Congresses. I. Rañada, A. F., 1939- II. Grupo Interuniversitario de Física Teórica. III. Title. IV. Series. QC20.7.N6155 1978 530.1'5 79-13997



Printing and binding: Beltz Offsetdruck, Hemsbach/Bergstr.
2153/3140-543210

## FOREWORD

This volume contains the Lectures delivered at the IX GIFT* International Seminar on Theoretical Physics which was held at Jaca, Huesca (Spain) in June 1978 on the general subject "Nonlinear Problems in Theoretical Physics". The lectures by F. Calogero and A. Degasperis were presented by D. Levi. The two seminars on nonlinear plasmas were delivered by C. Montes and M.R. Feix.

The Seminar was supported financially by the Instituto de Estudios Nucleares (Institute for Nuclear Studies), Madrid and took place at the Residence of the University of Zaragoza. We wish to express our thanks to both Institutions for their support. All the members of the Theoretical Physics group of the Universidad Complutense, Madrid, participated actively in the organization.

Madrid, February 1979

Antonio F. Rañada

* Grupo Interuniversitario de Física Teórica

IX G.I.F.T. Seminar

List of Assistants:

1) A.O. Barut
2) I. Bialynicki-Birula
3) C. Dafermos
4) W.A. Strauss
5) M.R. Feix
6) C. Montes
7) A.F. Rañada
8) A. Galindo
9) R. de la Llave
10) H. Velazquez
11) V. Iordachescu
12) F. Moreno
13) E. Rodriguez
14) G. Sierra
15) R. Mondaini
16) C. Bona
17) M. Jaulent
18) F. Magri
19) P.C. Sabatier
20) M. Sabatier
21) J. Leon
22) X. Llovet
23) E. Verdaguer
24) J. Sesma
25) L. Mas
26) M. Lorente
27) L. Poleti
28) L. Vazquez
29) L. Martínez
30) L.J. Boya
31) J. Gutierrez
32) L. García
33) G. García
34) J.F. Cariñena
35) J. Mateos
36) M. Santander
37) V. Delgado
38) A. Lopez
39) K. Kraus
40) J.E. Ribeiro
41) W. Heidenreich
42) W. Lucke
43) C.A. Dominguez
44) M. Buttiker
45) W.H. Steeb
46) A. Villani
47) M.F. Rañada
48) M.A. Rodriguez
49) D. Levi
50) A. Hirshfeld
51) M. Boiti
52) F.X. Fustero
53) R. Lapiedra
54) A. Chamorro
55) J. Ritter
56) M.A. Iglesias
57) L. Abellanas
58) J. Uson
59) J. Julve
60) F. Guil
61) M. Ramón
62) J. Chinea
63) C. Gomez
64) A. Alvarez
65) F. Carbajo
66) J. Ibañez
67) Mrs. Ibañez
68) A. González-Arroyo
69) C. Alvarez
70) I. Mellado
71) J. Gomis

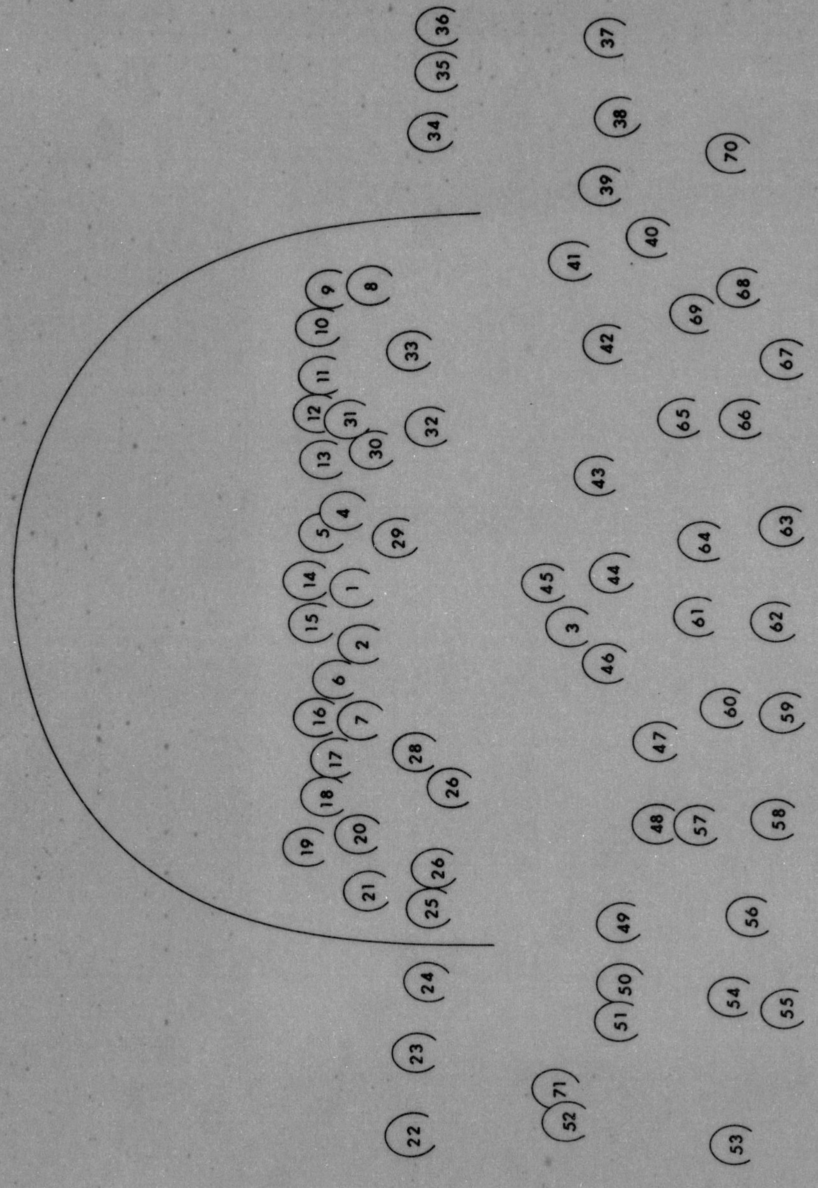

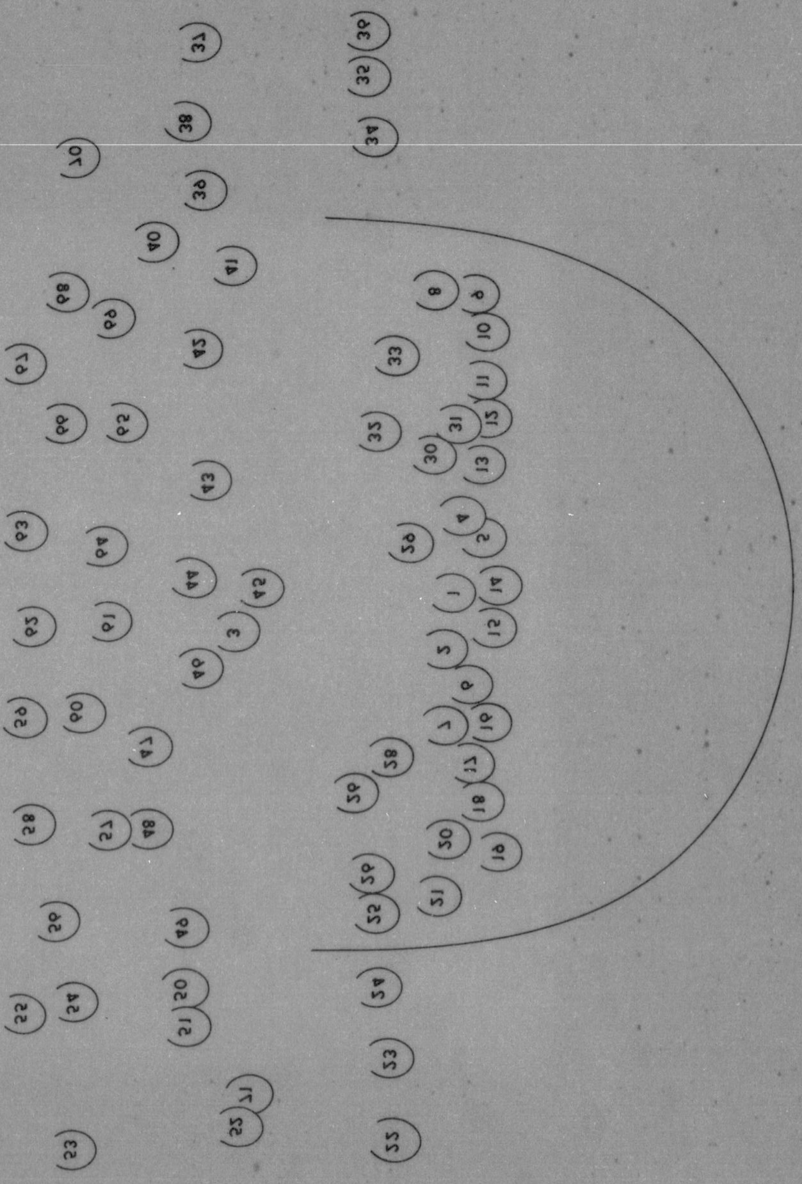

UNIVERSIDAD
DE
ZARAGOZA

# INDEX

NONLINEAR PROBLEMS IN THEORETICAL PHYSICS

IX INTERNATIONAL SEMINAR ON THEORETICAL PHYSICS

Jaca, June 5-10, 1978

## List of Participants

| | |
|---|---|
| Abellanas, L. | (Madrid, Complutense) |
| Alvarez, A. | (JEN, Madrid) |
| Alvarez, L. | (Madrid, Autónoma) |
| Arechalde, J.V. | (Madrid, Autónoma) |
| Barut, A.O. (Lecturer) | (Boulder, Colorado) |
| Berestycki, H. | (Paris) |
| Bialynicki-Birula, I. (Lecturer) | (Warsaw) |
| Bona, C. | (Barcelona Autónoma) |
| Boiti, M. | (Lecce) |
| Buttiker, M. | (Basel) |
| Carbajo, F.J. | (Madrid, Autónoma) |
| Cariñena, J.F. | (Zaragoza) |
| Chamorro, A. | (Bilbao) |
| Chinea, J. | (Madrid, Complutense) |
| Dafermos, C.M. (Lecturer) | (Brown Univ. Providence) |
| De la Llave, R. | (Madrid, Complutense) |
| Delgado, V. | (JEN, Madrid) |
| Dominguez, C.A. | (CIEA del IPN, México) |
| Feix, M.R. (Lecturer) | (Orléans) |
| Fernández, J.C. | (Nice) |
| Fernández Alvarez-E.,R. | (Madrid, Complutense) |
| Rañada, A.F. (Director of G.I.F.T. and the Organizer of the Seminar) | (Madrid, Complutense) |
| Fernández-Rañada, M. | (Zaragoza) |
| Fijalkow, E. | (Orléans) |
| Fustero, F.X. | (Barcelona, Autónoma) |
| Galindo, A. | (Madrid, Complutense) |
| García, A. | (UNAM, México) |
| García, G. | (Madrid, Complutense) |
| García, L. | (Madrid, Complutense) |
| Gómez, C. | (Madrid, Autónoma) |
| Gomis, J. | (Barcelona, Central) |
| González-Arroyo, A. | (Madrid, Autónoma) |
| González, F. | (CSIC, Madrid) |
| Guil, F. | (Madrid, Complutense) |
| Gutierrez, J. | (CSIC, Madrid) |
| Hirshfeld, A.C. | (Dortmund) |
| Ibañez, J. | (Madrid, Autónoma) |
| Iordachescu, V. | (Bures-sur-Yvette) |
| Heidenreich, W. | (München) |

| | | | |
|---|---|---|---|
| Jaulent, M. | (Montpellier) | Munier, A. | (Villeneuve-St-Georges) |
| Julve, J. | (CSIC,Madrid) | Pempinelli, F. | (Lecce) |
| Krauss, K. | (Wurzburg) | Pons, J.M. | (Barcelona, Central) |
| Lapiedra, R. | (Barcelona, Central) | Ramón, M. | (Madrid, Complutense) |
| Leon, J. | (Montpellier) | Ribeiro, J.E. | (Lisboa) |
| Levi, D. (Lecturer) | (Roma) | Ritter, J. | (King's College, London) |
| Leyland, P. | (Marseille) | Rodriguez, E. | (Madrid, Complutense) |
| Lions, P.L. | (ENS, Paris) | Rodriguez, M.A. | (Madrid, Complutense) |
| López, A. | (JEN, Madrid) | Sabatier, P.C. | (Montpellier) |
| Lorente, M. | (Madrid, Complutense) | Sánchez-Dehesa, J. | (Granada) |
| Lucke, W. | (Clausthal-Zellerfeld) | Santander, M. | (Valladolid) |
| Llovet, J. | (Barcelona, Autónoma) | Sesma, J. | (Zaragoza) |
| Magri, F. | (Milano) | Sierra, G. | (Madrid, Complutense) |
| Martín, J. | (Madrid, Autónoma) | Steeb, W.H. | (Paderborn) |
| Martínez, L. | (Madrid, Complutense) | Strauss, W.A. | (Brown Univ., Providence) |
| Mateos, J. | (Salamanca) | Usón, J.M. | (Madrid, Complutense) |
| Mas, L. | (Granada) | Vázquez, L. | (Madrid, Complutense) |
| Mellado, I. | (JEN, Madrid) | Velázquez, H. | (Tübingen) |
| Mondaini, R. | (ICTP, Trieste and Rio de Janeiro) | Verdaguer, E. | (Barcelona, Autónoma) |
| Montes, C. (Lecturer) | (Nice) | Villani, A. | (Sao Paulo) |
| Moreno, F. | (Madrid, Complutense) | | |

# NONLINEAR PROBLEMS IN CLASSICAL AND QUANTUMELECTRODYNAMICS

A.O. Barut

The University of Colorado, Boulder, Colorado 80309

Table of Contents

# NONLINEAR PROBLEMS IN CLASSICAL AND QUANTUMELECTRODYNAMICS

A. O. Barut
The University of Colorado, Boulder, Colorado 80309

## I. INTRODUCTION

We present here a discussions of the non-linear problems arising due to self-field of the electron, both in classical and quantum electrodynamics. Because of some shortcomings of the conventional quantumelectrodynamics |1| an attempt has been made to carry over the nonperturbative radiation reaction theory of classical electrodynamics to quantum theory. The goal is to have an equation for the radiating and self-interacting electron as a whole, in other words,an equation for the final "dressed" electron. In addition the theory and renormalization terms are all finite. Each particle is described by a single wave function $\psi(x)$ moving under the influence of the self-field as well as the field of all other particles. In particular, we discuss the completely covariant two-body equations in some detail, and point out to some new remarkable solutions of the non-linear equations: These are the resonance states in the two-body problem due to the interaction of the anomalous magnetic moment of the particle which become very strong at small distances.

## II. CLASSICAL RELATIVISTIC ELECTRON THEORY

The motion of charged particles are not governed by the simple set of Newton's equations as one usually assumes in the theory of dynamical systems, but by rather complicated non-linear equations involving even third order of derivatives. To see this we begin with Lorentz's fundamental postulates of the electron theory of matter:

(i) Matter consists of a number of charged particles moving under the influence of the electromagnetic field produced by all charged particles. The equation of motion of the $i^{th}$ charged particle is given by

$$m \ddot{Z}_{\mu}^{(i)} = e F_{\mu\nu}(x) \dot{Z}^{\nu(i)}\Big|_{X=Z^{(i)}} \quad , \tag{1}$$

where $Z_{\mu}(S)$ is the world-line of the particle in the Minkowski space $M^4$ in terms of an invariant time parameter S (e.g. proper time) - the derivatives are with respect to S, and $F_{\mu\nu}$ is the total electromagnetic field.

(ii) The total electromagnetic field $F_{\mu\nu}$ obeys Maxwell's equations

$$F_{\mu\nu}{}^{,\nu}(x) = j_\mu(x) \quad , \tag{2}$$

where $j_\mu(x)$ is the total current of all the charges. For point charges we have

$$j_\mu(x) = \sum_k e^{(k)} \dot{Z}_\mu^{(k)} \, \delta(x - Z_\mu^{(k)}) \tag{3}$$

We have in principle a closed system of equations if we have in addition some model of matter telling us how many charged particles there are.

These equations taken together give for each particle i a highly nonlinear equation because due to the term k = i in (3), $F_{\mu\nu}(x)$ in (1) depends nonlinearly on $Z^{(i)}$. This is the socalled <u>self-field</u> of the $i\underline{th}$ particle. Actually this term is even infinite at $X = Z^{(i)}$ due to the factor $\delta(X-Z^{(i)})$. In practice this infinite term does not cause as much trouble as it should-one simply drops such terms in first approximation. The reason for this is that a major part of the self-field is already taken into account as the inertia or mass of the particle on the left hand side of eq.(i): in other words, the mass m in (1) is the socalled renormalized mass $m_R$ as I shall explain in more detail. Unfortunately not the whole of the self-field is an inertial term in the presence of external forces. Otherwise the whole electrodynamics would be a closed and consistent theory without infinites. For a single particle, it is true by definition, that all the self-field is in the form of an inertial term because then the equation is $m_R \ddot{Z}_\mu = 0$. But the presence of other particles <u>modifies</u> the contribution of the self-field to an inertial term $m_R \ddot{Z}$. And this is really the whole story and problem of electrodynamics, classical or quantummechanical: How much of the self-field is inertia?. After the inertial term has been subtracted, the remainnder gives rise to observable effects which we call radiative phenomena like anomalous magnetic moment, Lamb shift, etc. I will now show first how this is done in classical electrodynamics, and the existence of nonlinear radiative phenomena like anomalous magnetic moment and Lamb shift even in classical mechanics.

Let us separate in Eq.(1) the self field term:

$$m_o \ddot{Z}_\mu = e\lambda F_{\mu\nu}^{ext}(x)\, \dot{Z}^\nu + eF_{\mu\nu}^{self}(x)\, \dot{Z}^\nu \Big|_{X=Z} \quad , \tag{4}$$

where I have introduced a parameter $\lambda(\lambda=1)$ in order to study the limit $\lambda \to 0$ for a free particle. The first term on the right hand side of (4) is finite, but the second term becomes infinite at X=Z. By various procedures one can however study the structure of this term |2|. The result is as follows. The self-field term in (4) can be written, using (3), as a sum of two terms

$$- \lim_{\varepsilon \to 0} \frac{e^2}{2\varepsilon} \ddot{Z}_\mu + \frac{2}{3} e^2 (\dddot{Z}_\mu + \dot{Z}_\mu \ddot{Z}^2) \ . \tag{5}$$

Here $Z_\mu$ depends on $\lambda$ as well, $Z_\mu = Z_\mu(S,\lambda)$. The first term is an inertial part which we write as $-\delta m \ddot{Z}_\mu$ and bring it to the left hand side of (4). We shall now "renormalize" eq.(4) such that for $\lambda \to 0$ we have the free particle eq. $m_R \ddot{Z}_\mu = 0$. The renormalization procedure is not unambiguous: we have to know to what form we want to bring our equations. The above requirement for $\lambda \to 0$ gives us the following final equation

$$m_R \ddot{Z}_\mu = e\lambda F^{\text{ext.}}_{\mu\nu}(x)\, \dot{Z}^\nu + \frac{2}{3} e^2 (\dddot{Z}_\mu + \dot{Z}_\mu \ddot{Z}^2) - \left[\frac{2}{3} e^2 (\dddot{Z}_\mu + \dot{Z}_\mu \ddot{Z}^2)\right]_{\lambda=0} , \tag{6}$$

$$m_R = m_o + \delta m \ .$$

Had we not subtacted the last term, a "free" particle ($\lambda = 0$) would be governed by a complicated equation, and that is not how mass is defined. Also, eq.(6), shows without the last term the peculiar phenomena of preacceleration and socalled run away solutions |3| which have bothered a lot of people up to present time. The last term in (6) eliminates these problems.

The nonlinear term in (6) has all the correct physical and mathematical properties:

1) $\Gamma_\mu \dot{Z}^\mu = 0$, where $\Gamma_\mu = C\,(\dddot{Z}_\mu + \dot{Z}_\mu \ddot{Z}^2)$,

2) It gives correct radiation formula and energy balance.

3) It is a non-perturbative exact result .

It has moreover, the physical interpretation as Lamb-shift and anomalous magnetic moment. These can be seen by considering external Coulomb or magnetic fields and evaluating iteratively the effect of the radiation reaction term |4|.

The classical theory can be extended to particles with spin |5|. The spin variables are best described today using quantities forming a Grassman algebra |6|. The main result, except for additional terms, is the same type non-linear behaviour radiation term as in eq.(6).

Some solutions of the radiative equations with spin are known |7|. They exhibit much of the typical erratic behavior of the trajectory around an average trajectory which we know from the Dirac equation as "zitterbewegung". Conversely, the classical limit of the Dirac equation is not a spinless particle, but a particle with a classical spin. Thus the spin of the electron must be an essential feature of the structure of the electron (not just an inessential addition).

## III. QUANTUM THEORY OF SELF-INTERACTION

We see thus that the electron's equation of motion is fundamentally non-linear. When we go over to quantum mechanics we do not quantize the "radiating, self-interacting electron" but first the free electron. Let us compare the classical and quantum equations parallely:

$$m_o \ddot{Z}_\mu = e\lambda F^{ext.}_{\mu\nu}(x=Z)\dot{Z}^\nu + e F^{self}_{\mu\nu}(x=Z)\dot{Z}^\nu$$

$$(-i\gamma^\mu \partial_\mu - m)\psi = e\gamma^\mu A(x)\psi(x) + ? \quad , \tag{7}$$

or, non-relativistically and for $\tilde{A} = 0$,

$$(i\hbar \frac{\partial}{\partial t} - \frac{\hbar^2}{2m}\Delta)\psi = U\psi + ? \quad . \tag{8}$$

We see that in the standard wave mechanics the nonlinear terms coming from self-field have been omitted, and a renormalized mass have been used. But this is only an approximation. Hence we wish now to complete the wave equations by the inclussion of the self-field terms.

Nonlinear terms have been added to (7) and (8) in order to have soliton-like solutions |8|, |9|. I should like to discuss here the non-linear terms in the standard theory, thus without introducing any new parameters.

We consider the basic framework of Lorentz, eqs.(1)-(3), but when the electron is described by a Dirac field $\psi(x)$:

$$(-i\gamma^\mu\partial_\mu - m_o)\psi = e\gamma^\mu\psi(x)A_\mu(x) \tag{9}$$

$$F_{\mu\nu}{}^{,\nu}(x) = j_\mu(x) = e\bar{\psi}(x)\gamma_\mu\psi(x) \quad . \tag{10}$$

In the gauge $A^\mu{}_{,\mu} = 0$, we can eliminate $A_\mu(x)$ from these equations and obtain the non-linear integrodifferential equations

$$(-i\gamma^\mu\partial_\mu - m_o)\,\psi(x) = e^2\gamma^\mu\psi(x)\int dy D(x-y)\bar\psi(y)\gamma_\mu\psi(y)\ .$$

The choice of the Green's function is not unique. Knowing the particle interpretation of the negative energy states in the hole theory, we may choose the causal Green's function D.

Again, as in the classical case, this equation is not yet complete: we must make sure that in the absence of external interactions the particle obeys the equation

$$(-i\gamma^\mu\partial_\mu - m_R)\psi(x) = 0\ . \tag{11}$$

For more than one particle we may either introduce several fields $\psi_1, \psi_2, \dots,$ or consider nonoverlapping localized solutions of one field. In the former case, we obtain the coupled set of equations:

$$\begin{aligned}(-i\gamma^\mu\partial_\mu - m_o^{(1)})\psi_1(x) &= e_1^2\gamma^\mu\psi_1(x)\int dy\ D(x-y)\ \bar\psi_1(y)\ \gamma_\mu\psi_1(y)\\ &+ e_1e_2\gamma^\mu\psi_1(x)\int dy D(x-y)\bar\psi_2(y)\gamma^\mu\psi_2(y) + \dots\end{aligned} \tag{12}$$

and similar equations for $\psi_2, \psi_3, \dots$ Here the first term is the self-interaction, the others interactions of other particles. For a localized $\psi_2$, for example, the second term in the stationary state gives correctly the interaction potential as

$$\frac{1}{r} + \frac{1}{r^2} + \dots\ ,$$

so that the Coulomb potential is modified at small distances, as it should be. Now let us write eq.(12) as

$$(-i\gamma^\mu\partial_\mu - m_o - \gamma^\mu\lambda A_\mu^{ext})\psi(x) = e^2\gamma^\mu\psi(x)\int dy\ D(x-y)\bar\psi(y)\gamma_\mu\psi(y)\ , \tag{13}$$

where again $\lambda$ is a parameter of the external potential. Let $\overset{o}{\psi}_\lambda(x)$ be a solution of the left hand side. In the iterative solution of the full equation, the right hand side gives, if we insert the solution of the homogeneous equation $\overset{o}{\psi}_\lambda(x)$,

$$I_\lambda(x) \equiv e^2\gamma^\mu\overset{o}{\psi}_\lambda(x)\int dy\ D(x-y)\ \bar{\overset{o}{\psi}}_\lambda(y)\ \gamma_\mu\overset{o}{\psi}_\lambda(y) \tag{14}$$

For $\lambda = 0$, we must have the free eq.(11). Hence $I_{\lambda=0}(x)$ must be used to renormalize the mass, and perhaps the field $\psi(x)$.

If we work with localized functions always, the theory, including renormalization procedure, is finite, and nonperturbative; it describes a dressed, radiating self-interacting particle.

We consider now in a bit detail the two-body coupled equations for $\psi$ and $\eta$:

$$
\begin{aligned}
(\gamma^\mu P_\mu - m_1)\psi &= e_1\gamma^\mu A_\mu^{(1)\text{self}}\,\psi + e_1\gamma^\mu A_\mu^{(2)}\,\psi \\
(\lambda^\mu P_\mu - m_2)\eta &= e_2\lambda^\mu A_\mu^{(2)\text{self}}\,\eta + e_2\lambda^\mu A_\mu^{(1)}\,\eta
\end{aligned}
$$

with

$$
\begin{aligned}
A_\mu^{(1)}(x) &= e_1 \int dy\, D(x-y)\, \bar\psi(y)\gamma_\mu\psi(y) \\
A_\mu^{(2)}(x) &= e_2 \int dy\, D(x-y)\, \bar\eta(y)\lambda_\mu\eta(y) \;.
\end{aligned}
\tag{15}
$$

We shall bring these equations into a manageable radial form using the ansatz

$$
\psi^{(i)}(x) = \sum_n \psi_n^{(i)}(\vec r)e^{-iE_m^{(i)}t} \;, \tag{16}
$$

where n labels the quantum numbers $(E_n,J,M,K)$ and

$$
\psi_n^{(i)}(\vec r) \equiv \begin{pmatrix} g_n^{(i)}(r)\,\Omega_n^{(i)}(\hat r) \\ if_n^{(i)}(r)\,\Omega_{n'}^{(i)}(\hat r) \end{pmatrix} \;, \quad \Omega_{J1M}(\hat r) \equiv \sum_{1m;\frac12\mu} C^{JM}\, Y_1^m(\hat r)\,\chi_\mu \;, \tag{17}
$$

with these substitutions, and

$$
D(x) = -\int \frac{d^4k}{(2\pi)^4}\frac{e^{-ikx}}{k^2+i\varepsilon} \tag{18}
$$

one obtains after much computations the coupled radial equations |10|

$$\frac{df_s}{dr} - \frac{K_s-1}{r} f_s + (E_s - m_1)g_s = \frac{1}{2\pi}\frac{e_1^2}{r^2} \sum_{\substack{E_s=E_m-E_n+E_r \\ n,m,1}} \int dr' \, V_{1E_nE_m}(r,r')$$

$$\times \left\{ g_r \left[ g_n'^* \, g_m' \, T_1^{nmsr} + f_n'^* \, f_m' \, T_1^{n'm'sr} \right] + f_r \left[ g_n'^* \, f_m' \, T_2^{nm'sr'} - f_n'^* \, g_m' \, T_2^{n'msr'} \right] \right\}$$

$$+ \frac{1}{2\pi}\frac{e_1e_2}{r^2} \sum_{E_s=E_m-E_n+E_r} \int dr' \, V_{1E_nE_m}(r,r') \left\{ g_r \left[ e_n'^* e_m' T_1^{nmsr} + d_n'^* d_m' T_1^{n'm'sr} \right] \right.$$

$$\left. + f_r \left[ e_n'^* \, d_m' \, T_2^{nm'sr'} - d_n'^* \, e_m' \, T_2^{n'msr'} \right] \right\}$$

$$\frac{dg_s}{dr} + \frac{K_s+1}{r} g_s - (E_s + m) f_s = -\frac{1}{2\pi}\frac{e_1^2}{r^2} \sum_{E_s=E_m-E_n+E_r} \int dr' \, V_{1E_nE_m}(r,r')$$

$$\left\{ f_r \left[ g_n'^* \, g_m' \, T_1^{nms'r'} + f_n'^* \, f_m' \, T_1^{n'm's'r'} \right] - g_r \left[ g_n'^* \, f_m' \, T_2^{nm's'r} - f_n'^* \, g_m' \, T_2^{n'ms'r} \right] \right.$$

$$- \frac{1}{2\pi}\frac{e_1e_2}{r^2} \sum_{E_s=E_m-E_n+E_r} \int dr' \, V_{1\,E_nE_m}(r,r') \left\{ f_r \left[ e_n'^* e_m' T_1^{nms'r'} + d_n'^* d_m' T_1^{n'm's'r'} \right] \right.$$

$$- g_r \left[ e_n'^* \, d_m' \, T_2^{nm's'r} - d_n'^* \, e_m' \, T_2^{n'ms'r} \right] ,$$

$$f' = f(r'), \quad g' = (r') , \quad \text{etc.} \tag{19}$$

There are two similar equations for $e_r$ and $d_r$.

Here the Kernels V are known integrals

$$V_{\tilde{1}\,E_n\,E_m}(r,r') = -\frac{1}{\pi} r^2 r'^2 \int_0^\infty \frac{k^2 dk \, j_{\tilde{1}}(kr) j_{\tilde{1}}(kr')}{(E_n - E_m)^2 - k^2 + i\varepsilon} \tag{20}$$

$$\xrightarrow{E_n \to E_m} \frac{1}{2}\frac{1}{2\tilde{1}+1} (r_< r_>)^{3/2} \left(\frac{r_<}{r_>}\right)^{\tilde{1}+\frac{1}{2}}$$

$T_1$ and $T_2$ are known functions of Clebsch-Gordon coefficients.

The terms on the right hand side are the various interaction and radiative potentials. To see these more clearly we specialize to the stationary 1s-state (of positronium, for example): $l=0$, $K=-1$, all $J = 1/2$, etc. Then

$$\frac{df}{dr} + \frac{2}{r} f + (E_1 - m_1)\, g = \frac{1}{2\pi} \frac{1}{r^2}\, g \int dr'\, V_{0E_1E_1}(r,r')$$

$$\times \left[e_1^2\, (g'g'^* + f'^*f') + e_1 e_2\, (e'^*e' + d'^*d')\right] - \frac{1}{2\pi} \frac{1}{r^2}\, f \int dr'\, V_{1E_1E_1}(r,r')$$

$$\times \left[e_1^2\, f'\, g' - e_1\, e_2\, d'\, e'\right] \tag{21}$$

and similarly the other equations. We shall refer to the first term as "electric", to the second term as "magnetic" potential, because they are multiplied by g and f, respectively.

One can see by explicit calculation that the contribution of the second particle, also in s-state, to the first gives an electric potential, in the limit $r \to \infty$,

$$\frac{e_1 e_2}{4\pi r}\, g + \ldots. \tag{22}$$

as we have noted earlier, and as $r \to 0$

$$\frac{Z\alpha m}{4\pi\gamma}\, g = \text{const.}\ g. \tag{22'}$$

The magnetic potential behaves like

$$\begin{aligned} &-\frac{1}{8m}\, \frac{2\gamma+1}{3}\, \frac{e_1 e_2}{r^2}\, f, && \text{as } r \to \infty \\ &-\frac{1}{3}\, \frac{(Z\alpha)^3\, m^2}{4\pi\gamma(2\gamma-1)}\, rf, && \text{as } r \to 0 \end{aligned} \tag{23}$$

If the second particle is heavy for example, we can use the Coulomb potential only and obtain

$$\frac{df}{dr} - \frac{K-1}{r} f + (E - m - \frac{e_1 e_2}{4\pi r}) g = \frac{1}{r^2} \frac{e_1^2}{2\pi} \left\{ g \int_0^\infty dr' \, V_{0EE} (r,r') \right.$$

$$\left. (g'^* g' + f'^* f') - \frac{4K^2}{4K^2 - 1} f \int_0^\infty dr' V_{1EE}(r,r') f' g' \right\}$$

$$= \frac{e_1^2}{2\pi} \frac{1}{r^2} \left[ F(r) f + G(r) g \right] \equiv V_m f + V_e g \; , \qquad (24)$$

similarly for the other equations.

Here we introduced the electric and magnetic form factors $G(r)$ and $F(r)$, respectively. The magnetic form factor $F(r)$ has the form (which we shall need later)

$$F(r) = C \int_0^\infty dr' \, V_{1EE}(r,r') \, f(r') g(r') = C'(1 - e^{-2r/r_0}(1 + \mathrm{polyn}(2r/r_0))) \; , \qquad (25)$$

thus starts from zero and approaches a constant for large distances, a behavior which we know from perturbation theory.

Let us compare this result with the Dirac equations for the electron with an anomalous magnetic moment $a$ in the Coulomb field

$$\frac{df}{dr} - \frac{K-1}{r} f + (E - m - \frac{e_1 e_2}{4\pi r}) g = a \frac{e_1 e_2}{2mr^2} f \qquad (26)$$

which is of course valid for $r \to \infty$, hence the identification of the anomalous magnetic moment interaction.

The anomalous magnetic moment has also an interaction with the self-field. Similarly, we have additional electric potentials, and, as we see from (21), a charge renormalization due to the term $(e_1^2/4\pi) \frac{1}{r} g$. But before using these values, we must renormalize the self-energy effects. In fact, from the integrals evaluated with the trial Coulomb type functions, for example we must subtract their values when $e_2 \to 0$, the free particle values.

## Other Remarkable Solutions of Nonlinear Equations

The Dirac equation in Coulomb field without the radiative terms on the right hand side, has the well-known discrete spectrum and the continium, the complete set of solutions is known. We get a hint for a new class of solutions with radiative terms corresponding to sharp resonances from eq.(24). Eliminating one of the radial functions f or g we can obtain a Schrödinger type eigenvalue equation |11|

$$\psi'' = (m^2 - E^2 + V_{eff}(E,r)\psi \quad , \qquad (27)$$

where

$$V_{eff} = \frac{K(K+1)}{r^2} + 2E\,U - U^2 + V_m^2 + 2V_m \frac{K}{r} - V_m'$$
$$+ \frac{1}{2}\,\frac{U'' - 2U'\frac{K}{r} - 2U'V_m}{m+E-U} + \frac{3}{4}\,\frac{U'^2}{(m+E-U)^2} \quad ,$$

$$U = V_1 + \frac{e_1 e_2}{4\pi r} \quad .$$

If we look at the shape of this potential for $V_1 = 0$, for example, we see that beside the usual Coulomb and centrifugal barrier, we have a rather large potential well at small distances and then the potential goes to $+\infty$ as $r \to 0$. For potentials of this type very high energy narrow resonances have been located numerically |12|. There is a simple intuitive picture for these resonances and they can be even calculated in semi-classical relativistic theory: They are the unstable states bound due to magnetic forces. They can be called the magnetic levels in contrast to the electric levels of the Coulomb problem. The corresponding two-body states in the case of $(e^+ - e^-)$ have been called superpositronium. They may be identified with the high energy narrow $\psi$-resonances in the $(e^+e^-)$ system |13|. We also predict similar resonances for $(e^-e^-)$ system.

Going back now to our nonlinear integral equation (24), we assume a localized solution of some size $r_0$, evaluate the form factor $F(r,r_0)$ with this form factor we plot $V_{eff}$ in (27) and look for a resonance with positive energy $E = E_{res}$. The form factor changes the form of the potential only at distances about $r \lesssim r_0$. The important size determining the position of the potential well is given by the value of the anomalous magnetic moment, i.e. $\alpha^2/m$. In principle one can iterate and look for a self-consistent solution: the wave function of the resonance must match fairly well the initial trial function.

## Some Related Problems

(1) The self-energy theory can be extended from electron to neutrino, in particular, the limit $e \to 0$, $m \to 0$ such that $e/m$ = const. implies an anomalous magnetic moment for neutrino. The problem of anomalous magnetic moment of $\nu$ has a long history. Here we wish to point out that the mechanism of magnetic resonances that we have discussed could also occur for the system $(e\nu)$ or $(\mu\nu)$, or even perhaps for $(\nu\nu)$. Furthermore, how much of the weak interactions can be simulated by the anoma

lous magnetic moment interactions is an open problem. The magnetic resonance phenomena can further be extended to three (or more) body systems, like (eee), (ee$\mu$), (ee$\nu$),etc.

(2) The self-energy interaction of the electron may give rise to a kind of "excited" state of the electron in the form of $\mu$. A heuristic calculation of $m_\mu/m_e = \frac{3}{2}\alpha^{-1} + 1$, on this basis has been given elsewhere |14|.

(3) As we have noted, the interval structure of the electron is essential for a closed complete and consistent theory of the electron and electrodynamics. This is also essential for the understanding of quantum principle. A calculation of the Planck's constant $\hbar$ in the quantum radiation energy $E = \hbar\omega$, using the structure of the electron (of the type given by "zitter - bewegung") and the radiation formula |15| may illustrates further the importance of nonlinear effects.

(4) It would be very interesting to find the classical relativistic limit of our self-field integrals (13) by the method of Rubinow and Keller |16| to see if the Lorentz-Dirac Eq.(6) is reproduced.

## REFERENCES

|1| P.A.M. Dirac, Proc. High Energy Physics Conf. Budapest 1977.

|2| P.A.M. Dirac, Proc. Roy. Soc. (London) A167, 148 (1938).
A.O. Barut, Phys. Rev. 10, 3335 (1974), and references therein.

|3| G.N. Plass, Rev. Mod. Phys. 33, 37 (1961).
F. Rohrlich, Classical Charged Particles, Addison-Wesley, Reading, 1965.

|4| A.O. Barut, Phys. Lett. 73B, 310 (1978).

|5| H.J. Bhabha and H.G. Corben, Proc. Roy. Soc. (London) A178, 273 (1941).

|6| F.A. Berezin and M.S. Marinov, Ann. of Phys. 104, 336 (1977).

|7| H.C. Corben, Phys. Rev. 121, 1833 (1961), and Nuovo Cim. 20, 529 (1961).

|8| A.F. Rañada, Intern. J. Phys. 16, 795 (1978).

|9| I. Bialynicki-Birula, J. Michielsky, Ann. Phys. 100, 62 (1976).

|10| A.O. Barut and J. Kraus, to be published, and Phys. Rev. 16D, 161 (1977).

|11| A.O. Barut and J. Kraus, J. Math. Phys. 17, 504 (1976).

|12| A.O. Barut and R. Raczka, Acta Physica Polon. (in press).

|13| A.O. Barut and J. Kraus, Phys. Lett. 59B, 175-178 (1975).

|14| See Ref. |4|.

|15| A.O. Barut, Z.f. Naturf. 33a, 993 (1978).

|16| S.I. Rubinow and J.B. Keller, Phys. Rev. 131, 2789 (1963).

# ON THE STABILITY OF SOLITONS

I. Bialynicki-Birula

Institute of Theoretical Physics, Warsaw University and Department of Physics, University of Pittsburgh, Pittsburgh.

Table of contents

# ON THE STABILITY OF SOLITONS

I. Bialynicki-Birula

Institute of Theoretical Physics, Warsaw University and Department of Physics, University of Pittsburgh, Pittsburgh.

## I. LIAPUNOV STABILITY

The notion of stability has several meanings both in mathematics and in physics. In the studies of solitons one usually uses the notion of stability derived from the notion of stable equilibrium in classical mechanics. A mechanical system is in a state of stable equilibrium if the energy of the system attains its minimum (at least locally). Small oscillations around the state of stable equilibrium can be decomposed into a superposition of harmonic modes of oscillations. The Hamiltonian describing these oscillations is obtained by retaining only quadratic terms in the expansion of the full Hamiltonian around the point of stable equilibrium. The point of stable equilibrium, therefore, is characterized by two conditions:

$$\delta E = 0 \tag{1}$$

$$\delta^2 E = 0 \tag{2}$$

Small oscillations around the point of stable equilibrium are stable in the sense of Liapunov. Alexander Liapunov, a Russian mathematician, introduced the notion of stability in the theory of differential equations. The Liapunov stability of a solution of a differential equation, or a set of differential equations, can be defined whenever the initial value problem is well posed. A solution $\phi_0(t)$ is called stable if small perturbations of the initial data at $t = 0$ lead to small changes of the solution for all $t > 0$.

For partial differential equations describing solitons (this term is used here to denote all localized solutions with finite energy) the notion of Liapunov stability must be modified. In most cases of interest we are dealing with equilibrium which is not stable but neutral with respect to some perturbations. Such perturbations represent "free motions" and are connected with the symmetries (translational, rotational, dilatational, etc.) of the system. Only those perturbations which do not represent "free motions" satisfy the Liapunov stability condition. The discussion of stability along these lines is given in every thorough review article on solitons (see for example |1|). In my lectures I shall introduce and explore a different notion of stability, which can be traced back to Henri Poincaré.

## II. POINCARE STABILITY

At the beginning of this century Poincaré has been working on the structure of the electron. He was the first to discover that it is difficult to reconcile relativistic invariance with the assumption that the dynamics of the electron is governed by purely electromagnetic forces. In a somewhat more modern language we can express his problem as follows. The fundamental physical quantity describing an extended system in a relativistic theory is the energy-momentum tensor $T_{\mu\nu}$, whose components are interpreted as the energy density ($T_{oo}$), the energy flux density ($T_{oi}$), or the momentum density, and the stress tensor ($T_{ij}$). For the electromagnetic field these components are:

$$T_{oo} = \frac{1}{2}(\vec{E}^2 + \vec{B}^2) \tag{3a}$$

$$T_{oi} = (\vec{E} \times \vec{B})_i \tag{3b}$$

$$T_{ij} = -E_iE_j - B_iB_j + \delta_{ij}\frac{1}{2}(\vec{E}^2 + \vec{B}^2) \tag{3c}$$

Let us consider the simplest case of a single, spherically symmetric, extended particle. In its rest frame, in view of the symmetry, $T_{oi} = 0$. Under a special Lorentz transformation (a boost in the x direction) the components of $T_{\mu\nu}$ transform in the following manner:

$$'T_{oo} = \frac{T_{oo} + \beta^2 T_{11}}{1 - \beta^2} \tag{4a}$$

$$'T_{o1} = \beta\frac{T_{oo} + T_{11}}{1 - \beta^2} \tag{4b}$$

where $\beta = v/c$ and nonprimed components refer to the rest frame. Integrating Eqs.(4) over 'x, 'y and 'z, we obtain

$$\int d^3{'x}\,'T_{oo} = \frac{1}{\sqrt{1-\beta^2}}\int d^3x\,(T_{oo} + \beta^2 T_{11}) \tag{5a}$$

$$\int d^3{'x}\,'T_{o1} = \frac{1}{\sqrt{1-\beta^2}}\int d^3x\,(T_{oo} + T_{11}) \tag{5b}$$

where we used the relation $d^3{'x} = d^3x\sqrt{1-\beta^2}$. The formulas (5) agree with relativistic transformations laws for the energy $E = \int d^3x\,T_{oo}$ and momentum $P_1 = \int d^3x\,T_{o1}$:

$$'E = E/\sqrt{1-\beta^2} \quad , \qquad P_1 = E/\sqrt{1-\beta^2} \tag{6}$$

only when the integral of $T_{11}$ vanishes. Changing the direction of the Lorentz boost, we can derive the same conditions for $T_{22}$ and $T_{33}$,

$$\int d^3x\, T_{ii} = 0 \quad (i = 1,2,3) \tag{7}$$

For purely electromagnetic forces conditions (7) can not be satisfied, because

$$T_{11} + T_{22} + T_{33} = \frac{1}{2}(\vec{E}^2 + \vec{B}^2) > 0 \tag{8}$$

In order to circumvent this difficulty, Poincaré suggested in his paper on the dynamics of the electron |2|, that one should introduce additional forces of a non-electromagnetic origin, the so called Poincaré pressures, acting at the surface of the electron. These forces should be so chosen that they give contributions to the stress tensor with the opposite sign and make all integrals of the diagonal components of $T_{ij}$ equal to zero. Poincaré's cohesive pressure compensates the explosive electromagnetic force and gives the electron a certain degree of stability. However, the stability is not perfect, because only the integrals of the stress tensor vanish; the stresses are balanced merely on the average. I shall call Eq.(7) the Poincaré stability condition. Every relativistic localized system with finite energy satisfies these conditions. An interesting and elegant example of such a system is the model of the electron described by the nonlinear electrodynamics of Born and Infeld |3|.

## III. PERFECT POINCARE STABILITY

The Poincaré stability condition (7) is, of course, satisfied if the system is totally stress-free, i.e. when all components of the stress tensor identically vanish

$$T_{ij} = 0 \tag{9}$$

I shall call Eq.(9) the perfect Poincaré stability (PPS) conditions. This strong version of the Poincaré condition is rooted in the general theory of relativity, similarly as the weak version (7) resulted from the special theory. Indeed the Einstein field equations, which form the mathematical basis of this theory, can be viewed as a fourdimensional version of the PPS condition (9). This is emphasized below by the appropriate choice of notation:

$$T^{\text{total}}_{\mu\nu} = T^{\text{matter}}_{\mu\nu} + T^{\text{space-time}}_{\mu\nu} = 0 \tag{10}$$

where

$$T^{\text{space-time}}_{\mu\nu} = \frac{c^4}{8\pi k}(R_{\mu\nu} - \frac{1}{2} g_{\mu\nu} R) \tag{11}$$

Thus, gravitating matter is perfectly stable since all its stresses are counterbalanced by the stresses induced in the space-time continuum. The rigidity of space-time is enormous; in the CGS units the coefficient $c^4/8\pi k$ is of the order of $10^{48}$. For example, to counterbalance the stress due to the electric field, the space need only bend 1 part per $10^{27}$ for every V/cm. In what follows I shall explore the significance of the PPS condition outside the scope of the theory of gravitation. Even though this condition is primarily meant to apply to relativistic theories, it also can be used without relativity.

## IV. LOGARITHMIC SCHRÖDINGER EQUATION

We have come across the PPS condition for the first time in our study with Jerzy Mycielski |4| of nonlinear wave mechanics. Searching for a nonlinear, Schrödinger-type equation, whose solutions could be viewed as bona fide wave functions, we found the equation with the logarithmic nonlinearity.

$$i\hbar\, \partial_t\, \psi(\vec{r},t) = \left[-\frac{\hbar^2}{2m}\Delta - b\ln(|\psi|^2 a^n)\right]\psi(\vec{r},t) \tag{12}$$

where a and b are constants and n is the dimensionality of the configuration space. Our equation is a member of the following class of nonlinear equations:

$$i\hbar\, \partial_t\, \psi(\vec{r},t) = \left[-\frac{\hbar^2}{2m}\Delta + U(\vec{r}) + F(|\psi|^2)\right]\psi(\vec{r},t) \tag{13}$$

The stress tensor for any equation of type (13), with a real function F, can be obtained from the hydrodynamical form of the nonlinear Schrödinger equation |4|:

$$\partial_t \rho + \nabla.\vec{j} = 0 \tag{14a}$$

$$m\,\partial_t \vec{j} + \nabla.\hat{T} + m\rho\nabla U = 0 \tag{14b}$$

where

$$\rho = |\psi|^2 \tag{15a}$$

$$\vec{j} = \frac{\hbar}{2mi}\psi^* \overleftrightarrow{\nabla}\,\psi \tag{15b}$$

$$T_{ij} = \frac{\hbar^2}{4m}(\partial_i\partial_j\psi^*.\psi + \psi^*.\partial_i\partial_j\psi - \partial_i\psi^*.\partial_j\psi - \partial_j\psi^*.\partial_i\psi) + \delta_{ij}\rho\left[F(\rho) - G(\rho)\right] \tag{15c}$$

and

$$G(\rho) = \frac{1}{\rho}\int_0^\rho d\rho'\, F(\rho') \tag{16}$$

The current $\vec{j}$ and the stress tensor $\hat{T}$ can also be expressed in terms of density and velocity:

$$\vec{j} = \rho\, \vec{v} \tag{17a}$$

$$T_{ij} = - \rho \left[\frac{\hbar^2}{4m} \partial_i \partial_j \ln \rho + \delta_{ij} (G - F)\right] + m \rho\, v_i v_j \tag{17b}$$

In the soliton rest frame the velocity $\vec{v}$ is zero and the PPS condition takes on the form:

$$\frac{\hbar^2}{4m} \partial_i \partial_j \ln \rho + \delta_{ij} (G - F) = 0 \tag{18}$$

One can show (cf. |4|) that the only solution of this set of equations is the logarithm for F and the Gaussian for $\rho$:

$$F = - b \ln (\rho A) \quad , \quad \rho = N\, e^{-2 m b\, |\vec{r} - \vec{n}|^2/\hbar^2}$$

where A,b and N are arbitrary constants.

In this way we arrive at the Schrödinger equation with the logarithmic nonlinearity. This is a nonlinear equation which in any number of dimensions possesses soliton-type solutions (we call them gaussons) obeying the PPS condition.

## V. RELATIVISTIC SYSTEMS

Now I shall turn to relativistic field theories. First, I would like to point out that here again one dimensional field theories are exceptional; every localized and static solution of relativistic field equations obeys PPS. This follows directly from the continuity equation for the energy-momentum tensor. In one space dimension the space component of this equation reads:

$$\partial_o T_{o1} + \partial_1 T_{11} = 0 \tag{19}$$

For static solutions $\partial_o T_{o1} = 0$ and therefore $\partial_1 T_{11} = 0$. For localized solutions this implies $T_{11} = 0$.

In more than one dimension the conditions $T_{ij} = 0$ impose, in general, very stringent restrictions and for simple theories they just can not be satisfied. One example has already been mentioned in these lectures: it is the pure electromagnetic theory. The stress tensor can not vanish there, because its trace is positive. Before I give examples of field theories in which the PPS condition can be satisfied, let me make a few remarks on the energy-momentum tensor.

In order to derive an expression for this tensor in a given relativistic theory we can use the standard prescription of the general theory of relativity, even though gravitation will play no role in our discussion. According to this prescription all ten components of the energy-momentum tensor are obtained as derivatives of the action with respect to the ten components of the metric tensor $g^{\mu\nu}$,

$$T_{\mu\nu}(x) = 2 \frac{\delta}{\delta g^{\mu\nu}(x)} \int d^4x \sqrt{-g} \;\; L \tag{20}$$

where g is the determinant of the matrix $g_{\mu\nu}$. The energy-momentum tensor defined by this formula, as a result of field equations will satisfy the continuity equation, if the action is invariant under general coordinate transformations |5|. For example, for the Maxwell field we obtain:

$$L = - \frac{1}{4} f_{\mu\lambda} \; f_{\nu\rho} \; g^{\mu\nu} \; g^{\lambda\rho} \tag{21}$$

$$\delta \int d^4x \sqrt{-g} \;\; L = \int d^4x \;\; \sqrt{-g} \; (- \frac{1}{2} f_{\mu\lambda} \; f_{\nu\rho} \; g^{\lambda\rho} + \frac{1}{8} f_{\lambda\sigma} f_{\rho\tau} g^{\lambda\rho} g^{\sigma\tau} g_{\mu\nu}) \; \delta g^{\mu\nu} \tag{22}$$

where we have used the following property of the determinant:

$$\delta g = - g \; g_{\mu\nu} \; \delta g^{\mu\nu} \tag{23}$$

Formulas (20)-(22) lead to

$$T_{\mu\nu} = f_{\mu\lambda} \; f^{\lambda}_{\nu} + \frac{1}{4} g_{\mu\nu} \; f_{\lambda\rho} \; f^{\lambda\rho} \tag{24}$$

The derivation of the energy-momentum tensor for a self-interacting scalar field $\phi$ is equally straightforward:

$$L = \partial_{\mu} \; \phi^{*} . \partial_{\nu} \; \phi \; g^{\mu\nu} - V(\phi^{*}\phi) \tag{25}$$

$$T_{\mu\nu} = \partial_{\mu} \; \phi^{*} \partial_{\nu} \; \phi + \partial_{\nu} \; \phi^{*} \partial_{\mu} \phi - g_{\mu\nu} \; (\partial_{\lambda} \; \phi^{*} \; \partial^{\lambda} \; \phi - V(\phi^{*} \; \phi)\,) \tag{26}$$

Here again one can check that the PPS condition in three dimensions has no nontrivial solutions.

The stress tensor can also be obtained from the Hamiltonian by differentiating it with respect to space components of the metric tensor. Since the time component and all mixed components of the metric will not be varied, I shall choose them in their Cartesian form

$$g_{oo} = 1 \quad , \quad g_{oi} = 0 \quad , \quad g_{ij} = - h_{ij} \tag{27}$$

where $h_{ij}$ is a positive definite,3x3 matrix. In evaluating the stress tensor from the Hamiltonian one must, however, be careful and before calculating the derivatives with respect to $g_{ij}$ one has to perform the Legendre transformation and express the Hamiltonian in terms of canonical variables. Only after this transformation the formula

$$T_{ij} = 2 \frac{\delta}{\delta h^{ij}} \int d^3x \sqrt{h}\ H(\vec{x}) \tag{28}$$

can be used. In the theory of the electromagnetic field the Legendre transformation takes us from $(\vec{B},\vec{E})$ variables to $(\vec{B},\vec{D})$ variables (see for example |6|)

$$D^k = \frac{\partial}{\partial E_k} h^{1/2} L = h^{1/2} h^{kl} E_l \tag{29}$$

and we obtain for the Hamiltonian:

$$H = \frac{1}{2} \int d^3x\ h^{-1/2} (D^i D^j h_{ij} + B^i B^j h_{ij}) \tag{30}$$

In the calculation of the stress tensor the following relations are useful:

$$\delta h_{ij} = - h_{ik} \delta h^{kl} h_{lj}\ ,\quad \delta h = - h\, h_{ij}\, \delta h^{ij},\quad \varepsilon_{ikm}\, \varepsilon_{jln}\, h^{kl}\, h^{mn} = 2h_{ij}\, h^{-1}$$

The result of this calculation coincides with the Maxwell stress tensor, when $h_{ij} = \delta_{ij}$. In the theory of the charged scalar field the Legendre transformation takes us from $(\phi,\dot{\phi})$ variables to $(\phi,\pi)$ variables,

$$\pi = \frac{\partial}{\partial \dot{\phi}} h^{1/2} L = h^{1/2} \dot{\phi}^* \tag{31}$$

and the Hamiltonian is:

$$H = \int d^3x \sqrt{h}\ (\pi^* \pi\, h^{-1} + \partial_i \phi^* \partial_j \phi h^{ij} + V(\phi^* \phi)) \tag{32}$$

The stress tensor obtained by taking the derivatives of (32) with respect to $h^{ij}$, in Cartesian coordinates in flat space is:

$$T_{ij} = \partial_i\ \phi^*\ \partial_j \phi + \partial_j\ \phi^*\ \partial_i\ \phi + \delta_{ij}\ (\pi^*\ \pi - \nabla\ \phi^* . \nabla\phi\ - V) \tag{33}$$

This expression coincides with the space part of (26).

The expression (28) for the stress tensor enables us to interpret the PPS condition as the condition that the energy is stationary with respect to all perturbations of the space metric $h^{ij}$. Does one also obtain the true minimum of the energy? As we shall see later, for several interesting systems, it is indeed so.

## VI. GAUGE FIELDS

By comparing formulas (3c) and (33) we infer that it is advantageous from the point of view of PPS to combine scalar fields and vector fields, because these two fields contribute to the stress tensor with opposite signs and a cancellation may take place. In two space dimensions such a cancellation may take place even with one vector field, i.e. with the electromagnetic field |7|. However, in three dimensions we need more than one vector field. Systems of vector fields (Yang-Mills fields) have been extensively studied in recent years and their coupling to scalar fields (Higgs fields) has also been thoroughly investigated. The ground is, therefore, well prepared to undertake the study of the PPS condition in the context of those gauge theories. I shall choose the best known example of such a theory: the SU(2) symmetric Yang-Mills-Higgs theory. There are three fields of each type in this model of the gauge theory and the Lagrangian with no self-interactions of the scalar fields is:

$$L = -\frac{1}{4} F^a_{\mu\nu} F^{a\mu\nu} + \frac{1}{2} (\nabla_\mu \phi)^a (\nabla^\mu \phi)^a \tag{34}$$

$$F^a_{\mu\nu} = \partial_\mu A^a_\nu - \partial_\nu A^a_\mu + \varepsilon_{abc} A^b_\mu A^c_\nu \tag{35}$$

$$(\nabla_\mu \phi)^a = \partial_\mu \phi^a + \varepsilon_{abc} A^b_\mu \phi^c \tag{36}$$

where the indices a,b,c take on values 1,2 and 3. The stress tensor can be obtained either by varying the action with respect to $g^{\mu\nu}$ or by varying the Hamiltonian with respect to $h^{ij}$. The second method leads to more interesting results. The Hamiltonian, written in terms of canonical variables is:

$$H = \frac{1}{2} \int d^3x h^{1/2} \{ D^i_a D^j_a h_{ij} + \frac{1}{2} B^a_{ij} B^a_{kl} h^{ik} h^{jl} + \pi^a \pi^a h^{-1} + (\nabla_i \phi)_a (\nabla_j \phi)_a h^{ij} \} \tag{37}$$

where

$$D^i_a = h^{1/2} h^{ij} F^a_{oj} , \quad B^a_{ik} = - F^a_{ik} \tag{38}$$

$$\pi^a = h^{1/2} (\nabla_o \phi)^a \tag{39}$$

In order to study the PPS condition it is convenient to rewrite the Hamiltonian in a different form, following the approach of Coleman et al. |8|, who investigated the lower energy bound in flat space.

$$H = \frac{1}{2}\int d^3x\, h^{1/2}\left\{\left[(D_a^i h^{-1/2} - \sin\alpha\, h^{ij}(\nabla_j\phi)_a)(D_a^k h^{-1/2} - \sin\alpha\, h^{kl}(\nabla_l\phi)_a) + \right.\right.$$

$$\left.\left. + (B_a^i h^{-1/2} - \cos\alpha\, h^{ij}(\nabla_j\phi)_a)(B_a^k h^{-1/2} - \cos\alpha\, h^{kl}(\nabla_l\phi)_a)\right] h^{ik} + \pi^a\pi^a h^{-1}\right\} +$$

$$+ \sin\alpha \int d^3x\, D_a^i (\nabla_i\phi)_a + \cos\alpha \int d^3x\, B_a^i (\nabla_i\phi)_a \tag{40}$$

where $B_a^i = \frac{1}{2}\varepsilon^{ijk} B_{jk}^a$. The first integrand in this formula is a sum of three positive quadratic forms. Therefore, the minimum of H as a functional of $h^{ij}$ is obtained when each term under the first integral vanishes, i.e. when (in flat space)

$$\vec{D}_a = \sin\alpha\,(\vec{\nabla}\phi)_a \quad , \quad \vec{B}_a = \cos\alpha\,(\vec{\nabla}\phi)_a \quad , \quad \pi^a = 0 \tag{41}$$

Eqs. (41) also guarantee that their solution satisfies the PPS condition, because the last two terms in the Hamiltonian do not depend on the metric tensor $h^{ij}$. All we have to check now is that Eqs. (41) also imply that $\vec{D}, \vec{B}, \phi$ and $\pi$ satisfy the field equations. This can be rather easily done. From Eqs. (41) it follows that

$$(\vec{\nabla}\,.\,\vec{\nabla}\,\phi)^a = 0 \tag{42}$$

since $\vec{B}$ is a covariant curl, $\vec{B}_a = (\vec{\nabla}\times\vec{A})_a$. Eq. (42) is the correct field equation for the scalar field, since $\pi = 0$. The equation for $\vec{D}$

$$(\vec{\nabla}.\vec{D})^a = 0 \tag{43}$$

follows from Eqs.(41) and (42). It is somewhat more complicated to check that the remaining field equation is satisfied:

$$(\vec{\nabla}\times\vec{B})_a - (\nabla_0\,\vec{D})_a = \vec{j}_a \tag{44}$$

To this end we notice that

$$(\vec{\nabla}\times\vec{B})_a = \cos\alpha\,(\vec{\nabla}\times\vec{\nabla}\phi)_a = -\cos\alpha\,\varepsilon_{abc}\,\vec{B}_b\phi_c = -\cos^2\alpha\,\varepsilon_{abc}\,(\vec{\nabla}\phi)_b\,\phi_c$$

$$(\nabla_0\vec{D})_a = \sin\alpha\,(\nabla_0\vec{\nabla}\phi)_a = \sin\alpha\,\varepsilon_{abc}\,\vec{D}_b\phi_c = \sin^2\alpha\,\varepsilon_{abc}\,(\vec{\nabla}\phi)_b\,\phi_c$$

$$\vec{j}_a = -\varepsilon_{abc}\,(\vec{\nabla}\phi)_b\,\phi_c \tag{45}$$

The solution of field equations determined by Eqs. (41) is known in the literature as a special case of the Julia-Zee dyon |9|. Coleman et al. have shown in |8| that dyons are stable under all perturbations of the field. I have shown above that they are also stable under all perturbations of the space metric. As a result of a change in the geometry of space, the energy of the dyon can only increase. Following |8| we can minimize the lower bound for the energy by a proper choice of the angle $\alpha$, which was so far arbitrary.

Similar results can be obtained in the space of four dimensions in the world of instantons. The Lagrangian again has the form given by Eq. (34), but all Greek indices now run from 0 to 4. The Hamiltonian has still the same form (37), but the number of components of the $\vec{D}$ field (four) is now different from that of the $\vec{B}$ field (six). Therefore, we can no longer use the expression (40) to find the lower bound of the energy. Instead, I shall write H, in four space dimensions, in the forms:

$$H = \frac{1}{2}\int d^4x\, h^{1/2}\Big[(D_a^i\, h^{-1/2} - h^{ij}\,(\nabla_j\phi)_a\,(D_a^k\, h^{-1/2} - h^{kl}\,(\nabla_l\phi)_a\, k_{ik} + \frac{1}{4}(h^{im}\, h^{jn}\, B_{mn}^a - \frac{1}{2}\,\varepsilon^{ijmn}\, B_{mn}^a\, h^{-1/2})\,(h^{kr}\, h^{lp}\, B_{rp}^a - \frac{1}{2}\,\varepsilon^{klrp}\, B_{rp}^a\,.\, h^{-1/2})\, h_{ik}\, h_{jl} + \pi_a\pi_a\Big] + \int d^4x\, D_a^i\,(\nabla_i\phi)_a + \frac{1}{4}\int d^4x\, \varepsilon^{ijkl}\, B_{ik}^a\, B_{jl}^a \tag{46}$$

The minimum of the energy with respect to $h^{ij}$ is attained when

$$D_a^i = (\nabla^i\phi)_a\,,\quad B^{aij} = \frac{1}{2}\,\varepsilon^{ijkl}\, B_{kl}^a\,,\quad \pi_a = 0 \tag{47}$$

Again one can check that the solutions of Eqs. (47) are also solutions of the field equations. In the simplest case, when $\vec{D} = 0 = \phi$, the solution is known as the instanton of Belavin et al. |10|. It is a self-dual solution of pure Yang-Mills equations in four space dimensions. In four space dimensions, unlike in three, the trace of the stress tensor of the Yang-Mills field is identically zero:

$$T_k^k = F_{kn}^a\, F^{ank} + \frac{1}{4}\,\delta_k^k\, F_{mn}\, F^{mn} = 0 \tag{48}$$

This is why we can have solutions of the pure Yang-Mills equations which satisfy the PPS condition. For the instanton solution we can analyze in full detail the second variation of the energy with respect to $h^{ij}$. For small deviations from flat geometry we obtain

$$\delta^2 H = \int d^4x\, T_{ij,kl}\, \delta h^{ij}\, \delta h^{kl} \tag{49}$$

$$T_{ij,kl} = B^2 (\delta_{ik} \delta_{jl} + \delta_{il} \delta_{jk} - \delta_{ij} \delta_{kl}) + B^a_{ik} B^a_{jl} + B^a_{il} B^a_{jk}$$

$$B^2 = \frac{1}{4} B^a_{ik} B^{ik}_a \tag{50}$$

The 16 x 16 matrix $T_{ik,jl}$ obeys the following Cayley-Hamilton equation:

$$\hat{T}^2 - 4 B^2 \hat{T} = 0 \tag{51}$$

from which it follows that the eigenvalues of T are 0 and 4 $B^2$. Variations of the metric which belong to the zero eigenvalue do not change the energy. One such variation is easily found: $h^{ij} = \lambda \delta^{ij}$. This is a well know result of the conformal invariance of the theory. The instanton can be uniformly stretched in all directions without a change in its energy. Conformal transformations are examples of "free motions" which were mentioned in Sec. I.

In both cases, in 3 and 4 dimensions, we were able to transform the Hamiltonian to the form:

H = (non-negative quadratic form) + (metric independent terms)

The h-independent terms depend only on the topology of the solutions, not on the Riemannian geometry of the space. They are known (modulo numerical multiplicative constants) as topological indices, Pontriagin indices or winding numbers. Similar decompositions of the Hamiltonian can be found in other theories. I do not know of any general rules that control this type of behavior.

Analogous results were found independently by Hosoya |11|, who has studied the action (instead of the Hamiltonian) for various nonlinear field theories with topological conservation laws and discovered the basic decomposition into the quadratic form plus no-h terms.

## REFERENCES


1. R. Jackiw, Rev. Mod. Phys. 49, 681 (1977).
2. H. Poincaré, Rediconti Circolo Mat. Palermo 21, 129 (1906).
3. M. Born and L. Infeld, Proc. Roy. Soc. 147, 552 (1934).
4. I. Bialynicki-Birula and J. Mycielski, Annals of Physics 100, 62 (1976).
5. A. Trautman, in Gravitation, L. Witten Ed., Wiley, New York, 1962.
6. I. Bialynicki-Birula and Z. Bialynicka-Birula, Quantum Electrodynamics, Pergamon Press, Oxford, 1975, p. 96.
7. H.J. de Vega and F.A. Shaposnik, Phys. Rev. D14, 1100 (1976).
8. S. Coleman, S. Parke, A. Neveu and C. M. Sommerfield, Phys. Rev. D15, 544 (1977)
9. M.K. Prasad and C.M. Sommerfield, Phys. Rev. Lett. 35, 760 (1975).
10. A.A. Belavin, A.M. Polyakov, A.S. Schwartz and Yu.S. Tupkin, Phys. Lett. 59B, 85 (1975).
11. A. Hosoya, Prog. Theor. Phys. 59, 1781 (1978).

# SPECTRAL TRANSFORM AND NONLINEAR EVOLUTION EQUATIONS

F. Calogero
Istituto di Fisica, Universita di Roma, 00185 Roma, Italy
Istituto Nazionale di Fisica Nucleare, Sezione di Roma

ABSTRACT

This is a terse introduction to the idea of the spectral transform method to solve nonlinear evolution equations.

The lecture notes by A. Degasperis printed after this paper, provide a complete treatment fo the spectral transform technique to solve nonlinear evolution equations, based on the approach that uses generalized wronskian relations as the main tool of analysis |1|, and presented in the (rather general) context of the class of nonlinear equations solvable via the spectral transform associated to the matrix Schrödinger eigenvalue problem (defined on the whole line, with "potentials" vanishing at infinity); and the subsequent paper by D. Levi surveys the analogous results for the discretized case. This "philosophical" introduction is instead confined to a terse description, in the simplest setting, of the idea of the spectral transform, and is therefore aimed only at those readers whose sole purpose is to understand the mere essence of this important mathematical idea.

The standard comparison of the spectral transform technique for solving nonlinear evolution equations is with the Fourier transform method to solve linear partial differential equations. Let us recall, in the simplest setting, how this works. Let

$$u_t(x,t) = -i\,\omega\left(-\frac{\partial}{\partial x}\right) u(x,t) \tag{1}$$

be a linear evolution equation, $\omega(z)$ being an entire function (say, a polynomial). The ("Cauchy") problem of interest is to compute $u(x,t)$ for $t > t_0$, given

$$u(x,t_0) = u_0(x) \tag{2}$$

We assume hereafter that all functions are defined over the whole real axis, $-\infty < x < +\infty$, and that they vanish ("sufficiently fast") as $x \to \pm\infty$; then the problem (1-2) is solved by the 3 formulae

$$\hat{u}(k,t_0) = \int_{-\infty}^{+\infty} dx\ \exp(-ikx)\ u(x,t_0) \tag{3}$$

$$\hat{u}(k,t) = \hat{u}(k,t_0)\ \exp\left[-i\omega(k)(t-t_0)\right] \tag{4}$$

$$u(x,t) = (2\pi)^{-1} \int_{-\infty}^{+\infty} dk\ \exp(ikx)\ \hat{u}(k,t) \tag{5}$$

Note that the applicability of this technique depends on the possibility to go bijectively from configuration space to Fourier space (via the inverse Fourier transform (3) and from Fourier space to configuration space (via the direct Fourier transform (5)), and on the fact that to the (complicated) time evolution (1) in configuration space there corresponds the (extremely simple) time evolution (4) in Fourier space. The relevance of these results to mathematical physics is of course enormous, because many physical (or, for that matter, natural) phenomena are, in some approximation, described by equations of type (2) (or their obvious generalizations); indeed it is here that originates the central rôle played by Fourier analysis in mathematical physics, or more generally in applied mathematics.

The spectral transform technique can be viewed as an analogous tool to solve and investigate (certain classes of) nonlinear evolution equations. Indeed the mechanism is very similar to that described above, but with one main difference; the mapping between configuration space and "spectral" space (the analog of Fourier space) is now nonlinear (although it is generally defined by linear equations). Let us illustrate the situation by a simple, and very recent, example |2|.

Consider the nonlinear partial differential equation

$$u_t + (12t)^{-1} \left[u_{xxx} - 6u_x u - 4xu_x - 2u\right] = 0 \quad , \qquad u \equiv u(x,t) \tag{6}$$

whose relevance is highlighted by its relation, via the simple change of dependent and independent variables

$$u(x,t) = (12t)^{2/3}\, q(y,t), \quad y = (12t)^{1/3} x \tag{7}$$

to the so-called "cylindrical KdV equation"

$$q_t + q_{yyy} - 6q_y q + (2t)^{-1} q = 0 \quad , \qquad q \equiv q(y,t) \tag{8}$$

that plays some rôle in plasma physics. Again one is interested in solving the ("Cauchy") problem of finding $u(x,t)$ for $t > t_0$ given

$$u(x,t_0) = u_0(x) \tag{9}$$

(we assume $t_0 > 0$, to avoid any problem associated with the singularity of the coefficient of the second term in (6) for $t = 0$).

Now associate to $u(x,t)$ a function $f(z,t)$ through the linear ("spectral") problem characterized by the second order ordinary differential equation

$$- \Psi_{xx} + \left[x + u(x,t)\right] \Psi = z\Psi \quad , \qquad \Psi \equiv \Psi(x,z,t) \tag{10}$$

letting $\Phi(x,z,t)$ and $F(x,z,t)$ be the solutions of this equation identified by the boundary conditions

$$\lim_{x \to +\infty} \{\Phi(x,z,t)/Ai(x-z\} = 1 \tag{11a}$$

$$\lim_{x \to -\infty} \{F(x,z,t)/\left[Bi(x-z) - i\,Ai(x-z)\right]\} = 1 \tag{11b}$$

(that are clearly consistent with the assumed asymptotic vanishing of $u(x,t)$ since Ai and Bi are the Airy functions |3|, so that $Ai(x-z)$ and $Bi(x-z)$ are two independent solutions of (10) with $u(x,t) \equiv 0$), and defining

$$f(z,t) = \pi \left[F_x(x,z,t)\ \Phi(x,z,t) - F(x,z,t)\ \Phi_x(x,z,t)\right] \tag{12}$$

The correspondence between the function $u(x,t)$, $-\infty < x < +\infty$, and its "<u>spectral transform</u>" $f(z,t)$, $-\infty < z < +\infty$, is biunivocal; the equations written above define uniquely $f(z,t)$ from $u(x,t)$; while the computation of $u(x,t)$ from $f(z,t)$ can be effected via the equations

$$M(x,x',t) = \int_{-\infty}^{+\infty} dz \left[|f(z,t)|^{-2} - 1\right] Ai(x-z)\ Ai(x'-z) \tag{13}$$

$$K(x,x',t) + M(x,x',t) + \int_x^{\infty} dx''\ K(x,x'',t)\ M(x'',x',t) = 0\ , \quad x' \geq x \tag{14}$$

$$u(x,t) = -2\, d\,K(x,x,t)/dx \tag{15}$$

(The first of these equations defines the kernel M; the second characterizes uniquely the function $K(x,x',t)$, being a Fredholm integral equation for its dependence on the second variable $x'$, while the dependence on the variables x and t is parametric, originating from the explicit appearance of these variables in (14); and the third of these equations defines u). Note that the relation between u and f is not linear (in contrast to the relation between u and its Fourier transform $\hat{u}$; see (3) and (5)), although the computation of f from u, as well as the computation of u from f, requires only the solution of linear equations (the "Schrödinger" equation (10), and the "Gel'fand-Levitan" equation (14)).

The solvability of the <u>nonlinear</u> evolution equation (6) is now implied by the (highly nontrivial!) fact that, if u evolves in time according to this equation, the corresponding time evolution of f is extremely simple, being given by the explicit formula

$$f(z,t) = f\left[z(t/t_0)^{1/3},\ t_0\right] \tag{16}$$

so that, to evaluate $u(x,t)$ from $u(x,t_0)$ (see (9)), one first evaluates $f(z,t_0)$ via (10-12)), then $f(z,t)$ (via (16)), and finally $u(x,t)$ (via (13-15)), thereby reducing the solution of the Cauchy problem for the <u>nonlinear</u> evolution equation (6) to a sequence of linear steps.

In addition to providing a technique for the solution of the Cauchy problem for (6), the association to $u(x,t)$ of its spectral transform $f(z,t)$ yields other qualitative and quantitative properties of the solutions of (6), for which the interested reader is referred to the original literature |2|.

As it has been described above, this technique appears extremely ad hoc. In fact, by appropriate choices of the spectral problem that institutes the connection between a function and its spectral transform, several classes of nonlinear evolution equations can be solved, including many of considerable practical interest; and many qualitative and quantitative properties of the solutions of these equations can be inferred, foremost among them being the appearance of the solitons, that are generally associated to the discrete part of the spectral transform, when it is present (this is not the case in the example we have for simplicity selected - although even in this case something analogous to the solitons could still be identified |2|). The interested reader will find a detailed treatment of such results in the following two papers. Here we conclude noting that the search for novel types of spectral transforms, and the identification of the corresponding classes of solvable evolution equations, constitutes today a very active and most promising research line.

REFERENCES

1. Besides the papers referred to in Degasperis'lecture notes, the following review papers are now available: F. Calogero, "Nonlinear evolution equations solvable by the inverse spectral transform", in Mathematical Problems in Theoretical Physics, edited by G.F. Dell'Antonio, S.Doplicher and G.Jona-Lasinio, Lecture Notes in Physics 80, Springer, 1978; F.Calogero and A.Degasperis, "Nonlinear evolution equations solvable by the inverse spectral transform associated to the matrix Schrödinger equation" in Solitons, edited by R.K.Bullough and P. J. Caudrey, Lecture Notes in Physics, Springer, 1979; A.Degasperis, "Solitons, Boomerons and Trappons", in Nonlinear Evolution Equations solvable by the Spectral Transform, Proceedings of a Symposium held at the Accademia dei Lincei in Rome (June 1977), edited by F. Calogero, Research Notes in Mathematics 26, Pitman, 1978; F. Calogero and A. Degasperis, "The Spectral Transform: a Tool to Solve and Investigate Nonlinear Evolution Equations", in Applied Inverse Problems, edited by P.C. Sabatier, Lecture Notes in Physics 85, Springer, 1978.

2. F. Calogero and A. Degasperis: "Inverse spectral problem for the one-dimensional Schrödinger equation with an additional linear potential", "Solution by the spectral transform method of a nonlinear evolution equation including as a special case the cylindrical KdV equation", "Conservation laws for a nonlinear evolution equation that includes as a special case the cylindrical KdV equation", Lett. Nuovo Cimento 23, 143, 150, 155 (1978).

3. See, for instance: M. Abramowitz and I.A. Stegun, Handbook of Mathematical Functions, Dover, New York, 1965, chapter 10.

# SPECTRAL TRANSFORM AND SOLVABILITY OF NONLINEAR EVOLUTION EQUATIONS

A. Degasperis
Istituto di Fisica - Universita di Roma 00185 Roma-Italy
Istituto di Fisica - Universita di Lecce
Istituto Nazionale di Fisica Nucleare, Sezione di Roma

Table of Contents

# SPECTRAL TRANSFORM AND SOLVABILITY OF NONLINEAR EVOLUTION EQUATIONS

## 1. THE SPECTRAL TRANSFORM

Many linêar evolution equations can be solved, and the properties of their solutions can be investigated, by means of the powerful method of the Fourier transformation. Indeed, today this method is inevitably part of the background of any scientist, from the pure mathematician to the engineer; of course, the reason is that its relevance and importance has gone well far beyond the theory of heat conduction, i.e. the subject of the original Fourier's treatise. In the last decade, after the important paper by GGKM |1|, it has become clear that, for certain classes of nonlinear evolution equations, the resolving method based on the Spectral Transform (ST) can be considered |2| as an extension of the Fourier analysis, to which it reduces by linearizing the nonlinear equation (with appropiate caution, see below). It is therefore reasonable to believe that the ST method opens highly interesting perspectives in applied, as well as in pure, mathematics; in any case, due to the fortunate fact that many important physical phenomena |3| are modeled by nonlinear partial differential equations which are "solvable" by the ST technique, this method is certainly a major breakthrough in mathematical physics.

By definition, a nonlinear evolution equation is solvable if its corresponding Cauchy problem can be solved by means of linear operations. The ST method consists on associating to a Solvable Nonlinear Evolution Equation (SNEE) a linear eigenvalue problem with respect to which the spectral transform of the solution of the SNEE is defined for any fixed value of time – in a way which will be detailed below. Then the spectral transform of the solution, and this is the crucial point, turns out to evolve with time according to a linear equation; of course, this scheme works only if one is able to invert the ST and, for this reason, this method has been more commonly referred to as the Inverse Spectral Transform (IST) method. It is true that the acronym IST has been introduced by AKNS |2| for the Inverse Scattering Transform because in their paper the associated linear eigenvalue problem was a wave scattering problem; however our redefinition seems more appropiate since this method can certainly apply also to SNEE's associated to linear operators which do not necessarely describe scattering processes |4|.

In order to find the class of SNEE's associated to a given linear eigenvalue problem |5|, several approaches have been proposed. The first systematic way to produce SNEE's is due to P. Lax |6| and is based on the isospectral time evolution of a linear operator; in fact, he proved that the linear Schroedinger operator $-\frac{d^2}{dx^2} + q(x,t)$ evolves according to an isospectral transformation if the "potential" q(x,t) is a solution of the well-known KdV equation (or of any of the so called higher KdV equations). A second approach, which is equivalent to the previous one and is based on the integrability conditions for a pair of partial differential equations, has been developed by AKNS |2|; within this powerful scheme, they discussed the important class of SNEE's associated to a non self-adjoint generalization of the Zakharov-Shabat linear problem. A third approach, introduced by F. Calogero, takes advantage of a generalization of the usual wronskian relations for solutions of a (generalized) Sturm-Liouville problem |7|. This last technique, showing a close analogy with the Fourier analysis, seems to be more appropiate to present in a unified and transparent way the main results of the theory (such as the Bäcklund transformations); its application to the multichannel Schrödinger problem will be the subject of these lectures. In the last few years, other approaches to nonlinear evolution equations have been developed within the framework of differential geometry |8|, Lie groups |9| and algebraic geometry |10|.

In the following we will focus upon dynamical systems described at time t by a multicomponent field Q(x,t) depending on one space coordinate x. Solvable nonlinear field equations with more than one space variable can also be obtained |11| by the method described below; however, this non trivial extension will not be discussed here.

Now we complete this preliminary discussion by recalling the 3 steps which are performed to solve the Cauchy problem for the much simpler case of a linear evolution equation. Because we are interested in the basic ideas, we focus here only on the simpler case of a scalar field, say q(x,t), satisfying the linear evolution equation

$$q_t(x,t) = -i\omega(-i\partial_x,t)\, q(x,t) \,, \quad q(x,t_0) = q_0(x) \tag{1.1}$$

where $\omega(z,t)$ is entire in z and the subscript t denotes partial differentiation with respect to time. First one goes over from the x-space to the k-space via the Fourier transformation

$$\hat{q}(k,t) = \int_{-\infty}^{+\infty} dx\, q(x,t)\, \exp(-ikx) \quad , \tag{1.2}$$

then the explicit time evolution of the Fourier transform $\hat{q}(k,t)$ is immediately obtained by integrating the following simple evolution equation implied by (1.1)

$$\hat{q}_t(k,t) = -i\omega(k,t)\,\hat{q}(k,t)\ , \quad \hat{q}(k,t_0) = \hat{q}_0(k) \tag{1.3}$$

namely,

$$\hat{q}(k,t) = \hat{q}_0(k)\,\exp\left[-i\int_{t_0}^{t} dt'\,\omega(k,t')\right] . \tag{1.4}$$

Finally the solution $q(x,t)$ at time $t \neq t_0$ is recovered by inverting the Fourier transformation

$$q(x,t) = (2\pi)^{-1}\int_{-\infty}^{+\infty} dk\ \hat{q}(k,t)\,\exp(i\,k\,x) . \tag{1.5}$$

Thus, given $q(x,t_0)$, one computes $\hat{q}(k,t_0)$ by (1.2), then $\hat{q}(k,t)$ by (1.4), then $q(x,t)$ by (1.5). By appropriately defining a spectral transformation which maps from the x-space to the k-space, we will show that this 3-step scheme applies also to a large class of nonlinear evolution equations. Additional important results, which have their simple counterparts in the linear case, will be reported in the last section.

We start with the definition of the ST of an $N \times N$ matrix $Q(x)$, depending on the real variable x; to simplify the introduction of the basic mathematical tools, we generally assume this matrix $Q(x)$ to be hermitian

$$Q(x) = Q^{+}(x) \tag{1.6}$$

although almost all results remain valid, with obvious modifications, even if this condition (1.6) does not hold. We emphasize that the extra effort needed to deal with a matrix-valued function of x, rather than with a scalar field, is not due to a mere sake of generality, but it is required in order to cover almost all interesting evolution equations discovered so far. In order to define the spectral transform of $Q(x)$, we associate to it the matrix Schrödinger operator

$$H = -d^2/dx^2 + Q(x) \tag{1.7}$$

acting on the $L^2$ vector valued functions of x. The operator (1.7) and its properties are well-known in potential scattering theory under appropriate assumptions on the potential matrix $Q(x)$. The following definition of ST can apply to all those matrix-valued functions which satisfy all the usual requirements of the theory of scattering |12|. However, for sake of simplicity, and because it is sufficient to cover the more interesting cases, we assume stronger conditions on $Q(x)$, namely that $Q(x)$ is finite valued (for real $x$) and that it vanishes asymptotically exponentially or faster, i.e. we assume that, for some positive $\varepsilon$

$$\lim_{x\to\pm\infty} \left[\exp(\varepsilon|x|)\ Q(x)\right] = 0 \quad . \tag{1.8}$$

These conditions together with (1.6) garantee that the operator (1.7) is self-adjoint and, therefore, that its spectrum is real.

The continuous part of the spectrum of the operator (1.7) is characterized by the matrix differential equation

$$\Psi_{xx}(x,k) = \left[Q(x) - k^2\right] \Psi(x,k) \tag{1.9}$$

together with the following boundary conditions

$$\Psi(x,k) \xrightarrow[x\to+\infty]{} \exp(-i\,k\,x) + R(k)\ \exp(i\,k\,x) \tag{1.10a}$$

$$\Psi(x,k) \xrightarrow[x\to-\infty]{} T(k)\ \exp(-i\,k\,x) \quad . \tag{1.10b}$$

Note that here $\Psi$ is the $N \times N$ matrix depending on $x$ and $k$ which is built out of $N$ independent vector solutions, used as its columns; the subscript $x$ means partial differentiation with respect to $x$ and, of course, $k$ is real and, by convention, positive. The asymptotic behaviour (1.10a,b) of the solution $\Psi$ is characterized by two $N \times N$ matrices: the "reflection coefficient" $R(k)$ and the "transmission coefficient" $T(k)$; $R(k)$ depends on the positive real variable $k$ and its values for negative $k$ can be obtained by hermitian conjugation

$$R(-k) = R^{+}(k) \ , \tag{1.11}$$

the validity of this formula being implied by the results given below.

The discrete part of the spectrum consists of a finite number of negative eigenvalues of the operator (1.7); the fact that the number of these eigenvalues is finite is a consequence of (1.8) (of course, this number may also be zero). Each eigenvalue is characterized by the positive number $p^{(j)}$ which enters into the eigenvalue equation

$$\psi_{xx}^{(j)}(x) = \left[Q(x) + p^{(j)^2}\right] \psi^{(j)}(x) \quad , \quad j = 1,2,..\ M \ . \tag{1.12}$$

Without any loss of generality, we assume the eigenvalues $-p^{(j)2}$ to be nondegenerate; in fact, the case of degenerate eigenvalues can be easily recovered by a suitable limiting process (see below). With this assumption, the (one column) vector solution $\psi^{(j)}(x)$ of the equation (1.2) satisfying the normalization condition

$$\int_{-\infty}^{+\infty} dx \left[\psi^{(j)}(x)\right]^{+} \psi^{(j)}(x) = 1 \quad , \quad j = 1,2,.. M \tag{1.13}$$

uniquely defines, through its asymptotic behaviour

$$\psi^{(j)}(x) \xrightarrow[x \to +\infty]{} c^{(j)} \exp(-p^{(j)}x) \quad , \quad j = 1,2,.. M \tag{1.14}$$

the vector $c^{(j)}$ corresponding to the eigenvalue $-p^{(j)2}$. For future convenience, we notice that for a large class of matrices $Q(x)$, $R(k)$ and $T(k)$ can be analytically continued in the variable k and are indeed meromorphic in the whole complex k-plane if the matrix $Q(x)$ vanishes asymptotically faster than exponentially, i.e. if for any real number $\alpha$

$$\lim_{x \to \pm\infty} \left[\exp(\alpha x)\; Q(x)\right] = 0 \quad . \tag{1.15}$$

In this case a definite connection exists between the continuous part and the discrete part of the spectrum,in fact the matrix $R(k)$ turns out to have M simple poles at the values $k^{(j)} = ip^{(j)}$, $j = 1,2.. M$, the corresponding residues being related to the vectors $c^{(j)}$ by the simple formula |12|

$$\lim_{k \to ip^{(j)}} \left\{\left[k - ip^{(j)}\right] R(k)\right\} = ic^{(j)}\, c^{(j)+} \tag{1.16}$$

where the dyadic notation has been used. In the following, it will be convenient to characterize the vector $c^{(j)}$ corresponding to the j-th discrete eigenvalue, with its modulus

$$\rho^{(j)} = c^{(j)+} c^{(j)} \quad , \quad j = 1,2,.. M \tag{1.17}$$

and its direction to which we associate the corresponding $N \times N$ projection matrix

$$P^{(j)} = (\rho^{(j)})^{-1}\, c^{(j)}\, c^{(j)+} \quad , \quad j = 1,2,.. M \tag{1.18}$$

satisfying the obvious conditions

$$P^{(j)2} = P^{(j)} \quad , \quad \mathrm{tr}\; P^{(j)} = 1 \;,\; j = 1,2,.. M\;. \tag{1.19}$$

In conclusion, we attach to each point of the discrete spectrum (if any) a positive number $p^{(j)}$ (the eigenvalue being $-p^{(j)2}$), a positive number $\rho^{(j)}$ and a $N \times N$ matrix $P^{(j)}$ projecting on a one-dimensional subspace.

The ST of the matrix valued function Q(x) is defined as the following collection of quantities

$$\mathrm{ST} : \{R(k),\ -\infty < k < +\infty,\ p^{(j)},\ \rho^{(j)},\ P^{(j)},\quad j = 1,2,\dots M\} . \tag{1.20}$$

The rationale of this definition will be clear in the following. Thus, the computation of the ST of a given matrix Q(x) requires the solution of the direct matrix Schrödinger problem, namely the determination of the matrix R(k) and the spectral parameters $p^{(j)}$, $\rho^{(j)}$ and $P^{(j)}$ through Eqs. (1.9), (1.10), (1.12), (1.13), (1.14), (1.17) and (1.18). Such a computation defines the mapping $Q(x) \longrightarrow \{R(k), p^{(j)}, \rho^{(j)}, P^{(j)}, \quad j = 1,2,.. M\}$. Two very important differences between this mapping (i.e. $Q(x) \to$ ST of Q(x)) and the familiar Fourier transformation should be inmediately underlined; first, the ST of Q(x) depends clearly in a nonlinear way on Q(x), and this nonlinearity is the essential property of the ST which makes possible to solve nonlinear evolution equations. The second difference refers to the natural splitting of the ST into two main parts, namely that one corresponding to the continuous spectrum of the Schrödinger operator (1.7) characterized by the matrix valued function R(k), and the other one corresponding to the discrete spectrum of the operator (1.7) characterized by the M pairs of positive numbers $p^{(j)}$ and $\rho^{(j)}$ together with their corresponding M hermitian projectors $P^{(j)}$. These two components of the ST are completely independent from each other and will play a very different role in the application of the ST method to the investigation of the time evolution of physical systems. Indeed, the very existence of the discrete component of the ST will be reflected in the existence of the so-called solitons.

The problem of finding the ST of a given Q(x) will be referred to as the direct problem. Its solution requires the integration of the $N \times N$ matrix Schrödinger equation. It is worthwhile to reformulate very briefly the direct problem in terms of the well-known integral equations of scattering theory |13|. For sake of simplicity, and also because we want to focus on the analogy between the Fourier transform and the ST, we will consider now only the continuous component of the ST, i.e. the reflection coefficient R(k). The starting point is the integral equation satisfied by the matrix solution of the differential equation (1.9), together with the boundary condition (1.10b)

$$\Psi(x,k) = T(k)\exp(-i\,k\,x) + \int_{-\infty}^{+\infty} dy\; g(x-y,k)\; Q(y)\; \Psi(y,k) \tag{1.21}$$

where g(z,k) is the appropriate Green function

$$g(z,k) = (1/k)\sin(kz)\,\theta(z) \tag{1.22}$$

$\theta(z)$ being the step function, i.e. $\theta(z) = 1$ for $z > 0$ and $\theta(z) = 0$ for $z < 0$. The asymptotic behaviour (1.10a) is then recovered from the integral equation (1.21) with the following identification

$$T(k) = 1 + (2ik)^{-1}\int_{-\infty}^{+\infty} dx\,\exp(ikx)\; Q(x)\; \Psi(x,k) \tag{1.23}$$

$$R(k) = (2\,i\,k)^{-1}\int_{-\infty}^{+\infty} dx\,\exp(-ikx)\; Q(x)\; \Psi(x,k)\,. \tag{1.24}$$

These integral representations of the reflection and transmission coefficients are not quite appropriate to discuss the problem of obtaining approximate expressions of T and R in terms of Q. In fact, because of the $k^{-1}$ factor in front of the integrals in (1.23) and (1.24), the first order approximation (obtained setting $\Psi(x,k) \simeq \exp(-ikx)$) would be certainly very poor in the long wave-length limit. It turns out to be more convenient to introduce the following matrix solution of the Schrödinger equation (1.9) |14|

$$\Phi(x,k) \equiv \Psi(x,k)\left[T(k)\right]^{-1} \tag{1.25}$$

which satisfies the integral equation

$$\Phi(x,k) = \exp(-ikx) + \int_{-\infty}^{+\infty} dy\; g(x-y,k)\; Q(y)\; \Phi(y,k) \tag{1.26}$$

and also to introduce, correspondingly, the N x N matrix valued function of two variables

$$\mathcal{R}\,(k',k) \equiv \int_{-\infty}^{+\infty} dx\,\exp(-ik'x)\; Q(x)\; \Phi(x,k) \tag{1.27}$$

which is itself the solution of the following integral equation

$$\mathcal{R}(k'k) = \hat{Q}(k'+k) + (2\pi)^{-1} \int_{-\infty}^{+\infty} dq \left[k^2 - (q - i\varepsilon)^2\right]^{-1} \hat{Q}(k'-q)\, \mathcal{R}(q,k) \tag{1.28}$$

$\hat{Q}(k)$ being the Fourier transform of $Q(x)$ (see Eq.(1.31)). This integral equation is derived from Eqs.(1.27) and (1.26) by using standard techniques of scattering theory, and is known as the Lippmann-Schwinger integral equation. Then the expression of the reflection and transmission matrices can be obtained in terms of the matrix $\mathcal{R}(k',k)$ from the definitions (1.25) and (1.27), and the integrals (1.23) and (1.24); they read

$$R(k) = \mathcal{R}(k,k) \left[2ik - \mathcal{R}(-k,k)\right]^{-1} \tag{1.29}$$

$$T(k) = 2ik \left[2ik - \mathcal{R}(-k,k)\right]^{-1} . \tag{1.30}$$

In the simple case with no discrete part of the spectrum the expression (1.29) and the integral equation (1.28) clarify the relationship between the Fourier transform of the matrix $Q(x)$

$$\hat{Q}(k) = \int_{-\infty}^{+\infty} dx \exp(-ikx)\, Q(x) \tag{1.31}$$

and the linear approximation of the ST of $Q(x)$

$$R(k) \simeq (2ik)^{-1} \hat{Q}(2k) \; ; \tag{1.32}$$

however, by setting into the expression (1.29) the linear approximation $\mathcal{R}(k',k) \simeq \hat{Q}(k'+k)$, one obtains the improved approximation of the ST of $Q(x)$

$$R(k) \simeq \hat{Q}(2k) \left[2ik - \hat{Q}(0)\right]^{-1} . \tag{1.33}$$

Therefore the linear integral equation (1.28) and the expression (1.29) are the basic equations not only to solve the direct problem, but also to investigate approximate expressions of the ST; this second point, however, will not be discussed any further.

In the same way as for the Fourier transformation, the applicability of the ST technique to solve evolution equations relies on the invertibility of the mapping which associates the set of quantities (1.20) to a matrix $Q(x)$. That the Schrödinger spectral problem defines a one-to-one correspondence between a large class of "potentials" $Q(x)$ and their corresponding ST: $\{R(k),\ -\infty < k < +\infty,\ p^{(j)},\ \rho^{(j)},\ P^{(j)},$ $j = 1,2,\ldots M\}$ is one of the most important result of the scattering theory, which has been discovered long time before its application to nonlinear evolution equations was recognized by GGKM. The solution of the inverse problem |15|, that is the problem of reconstructing the matrix $Q(x)$ from the knowledge of its spectral transform, is then obtained by the following procedure: i) compute the matrix

$$M(z) = (2\pi)^{-1} \int_{-\infty}^{+\infty} dk \exp(ikz) R(k) + \sum_{n=1}^{M} \rho^{(n)} P^{(n)} \exp(-p^{(n)} z) \tag{1.34}$$

then ii) solve the Fredholm matrix integral equation

$$K(x,x') + M(x+x') + \int_{x}^{+\infty} dy\, K(x,y)\, M(y+x') = 0 \quad , \quad x \leq x' \tag{1.35}$$

and finally iii) obtain $Q(x)$ by the simple formula

$$Q(x) = -2\, dK(x,x)/dx \, . \tag{1.36}$$

We notice that the integral equation (1.35), known as the Gel'fand-Levitan-Marchenko equation, is linear, and that the variable x enters in this equation only as a parameter. Of course, as for the mapping $Q(x) \rightarrow \{R(k), p^{(j)}, \rho^{(j)}, P^{(j)}\}$,also the inverse mapping $\{R(k), p^{(j)}, \rho^{(j)}, P^{(j)}\} \rightarrow Q(x)$, defined by Eqs. (1.34), (1.35) and (1.36), is nonlinear; the matrix $Q(x)$ will be then understood to be the Inverse Spectral Transform (IST) of $\{R(k), p^{(j)}, \rho^{(j)}, P^{(j)}\}$.

In analogy with what we are well used to do by means of the Fourier transformation, we are now in the position to reformulate problems concerning a matrix $Q(x)$ into the equivalent problems dealing with its ST and viceversa; or, in other words, to investigate certain properties of $Q(x)$ by looking at the corresponding properties of its ST, and viceversa. In the following, to be concise, we will refer to x-space or to k-space in the cases we will look at the $Q(x)$ or at its ST, respectively; which one of these two representation is more convenient depends, of course, on the problem being investigated. An important instance of this general rule will be given in the next Section by the nonlinear evolution equations, which are much more conveniently investigated in k-space than in x-space.

It is remarkable that a special class of matrix valued functions $Q(x)$ exists such that their ST has a known explicit analytic expression; these special $Q(x)$'s are characterized by a ST with vanishing reflection coefficient, to say of the type $\{R(k) = 0, p^{(j)}, \rho^{(j)}, P^{(j)}, j = 1,2,\ldots M\}$. Indeed, if only the discrete part of the ST is present, with M discrete eigenvalues, the kernel of the Gel'fand-Levitan-Marchenko equation is separable of rank M and this integral equation (1.35) reduces to an algebraic equation. It is an easy exercise to derive the simplest class characterized by just one discrete eigenvalue, namely by two positive real parameters p and $\rho$, and by a $N \times N$ one-dimensional projection matrix P; in fact, setting $R(k) = 0$, $M = 1$, $p^{(1)} = p$, $\rho^{(1)} = \rho$ and $P^{(1)} = P$ in Eq.(1.34) and solving the corresponding integral equation (1.35), one obtains the IST of $\{R(k) = 0, p, \rho, P\}$

$$Q(x) = - \{A/\cos h^2 \left[(x - \xi)/\lambda\right] \} P \tag{1.37}$$

where we have introduced the following new parameters

$$A = 2p^2 \quad , \; \lambda = p^{-1} \quad , \quad \xi = (2p)^{-1} \ln(\rho/2p) \tag{1.38}$$

in terms of the spectral parameters p and $\rho$, since they have a very transparent interpretation in x-space, namely A is the amplitude, $\lambda$ is the width and $\xi$ is the position of the (single) minimum. It is important to notice that the amplitude and the width are related to each other in a definite way, i.e. $A = 2/\lambda^2$, and that the matrix character of Q(x) is factorized in the x-independent projector P. For sake of completeness, we now report the expressions that other quantities, defined above, take in this particularly simple, but important, case:

$$K(x,x') = -2p\rho \exp\left[-p(x+x')\right] \left[2p + \rho \exp(-2px)\right]^{-1} P \tag{1.39}$$

$$\Phi(x,k) = \exp(-ikx) \{ 1 - \left[4ip^2/(k+ip)\right] \left[2p + \rho \exp(-2px)\right]^{-1} P \} \tag{1.40}$$

$$\mathcal{R}(k',k) = \pi\left[(k'^2-k^2)/(k+ip)\right]\exp\left[-i\xi(k'+k)\right] \{ \sinh\left[\pi(k'+k)/2p\right] \}^{-1} P \tag{1.41}$$

$$T(k) = 1 + \left[2ip/(k - ip)\right] P . \tag{1.42}$$

Let us now consider a one-parameter family of matrices Q(x,t) by introducing a parametric dependence on a new variable t, from now on taken as the time. Then, of course, all the corresponding quantities we previously defined will also depend on time. Thus, as Q(x,t) evolves in time, the solutions of the Schrödinger equations (1.9) and (1.12) evolve in time and therefore, through the definitions (1.10a), (1.12), (1.14), (1.17), (1.18) and (1.20), also the ST of Q(x,t) evolves in time: $\{R(k,t), p^{(j)}(t), \rho^{(j)}(t), P^{(j)}(t), j = 1,2,\dots M(t)\}$; of course, Eq.(1.10b) implies as well a t-dependence of the transmission coefficient T(k,t).

In analogy with the Fourier transformation and its application to solve linear partial differential equations (see (1.1), (1.5) and (1.4) ) it is natural now to investigate whether a time evolution exists such that, although possibly very complicate in x-space, is simple in k-space. The first result of this investigation is the striking discovery of a class of nonlinear partial differential equations whose solutions $Q(x,t)$ are such that their corresponding Spectral Transforms satisfy a linear ordinary differential equation. By definition, a nonlinear evolution equation of this class is solvable because its corresponding Cauchy problem can be solved by means of linear operations only. The way this is accomplished is schematically shown in the following "solvability diagram"

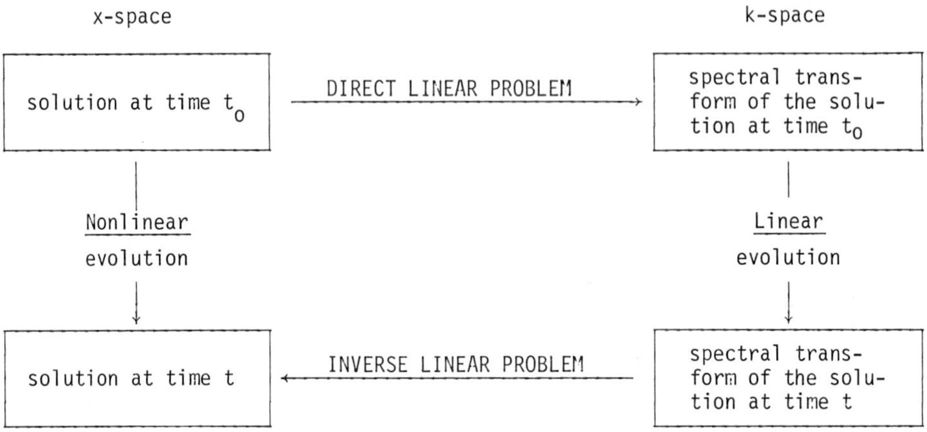


In order to derive, via the generalized wronskian relations technique, this class of SNEE, we need few more definitions which are tersely reported below. For a given matrix $Q(x)$, we introduce the following subsidiary Schrödinger spectral problem, which, for the continuous part of the spectrum is defined by the differential equation

$$\bar\Psi_{xx}(x,k) = \bar\Psi(x,k)\left[Q(x) - k^2\right] \quad , \qquad k \geq 0 \qquad (1.43)$$

the $N \times N$ matrix solution $\bar\Psi(x,k)$ being defined by the asymptotic conditions

$$\bar\Psi(x,k) \xrightarrow[x \to +\infty]{} \exp(-i\,k\,x) + \bar R(k)\exp(i\,k\,x) \qquad (1.44a)$$

$$\bar\Psi(x,k) \xrightarrow[x \to -\infty]{} \bar T(k)\exp(-i\,k\,x)\;; \qquad (1.44b)$$

(1.43) differs from (1.9) because now the matrix $Q$ on the r.h.s. acts from the right rather than from the left. The discrete part of the spectrum is characterized by the vector differential equation

$$\bar\psi^{(j)}_{xx}(x) = \bar\psi^{(j)}(x)\left[Q(x) + \bar p^{(j)2}\right] \quad , \quad \bar p^{(j)} > 0 \quad , \quad j = 1,2,\dots \bar M \qquad (1.45)$$

together with the normalization condition

$$\int_{-\infty}^{+\infty} dx\, \bar{\psi}^{(j)}(x)\, \bar{\psi}^{(j)^{+}}(x) = 1\;,\; j = 1,2,\ldots \bar{M}\;. \qquad (1.46)$$

Here the solution $\bar{\psi}^{(j)}(x)$ is a N-dimensional row vector valued function (therefore, in (1.46), the integrand is the usual scalar product), and its asymptotic behaviour, because of the condition (1.46), uniquely defines the row vector $\bar{c}^{(j)}$

$$\bar{\psi}^{(j)}(x) \xrightarrow[x \to +\infty]{} \bar{c}^{(j)}\exp(-\,\bar{p}^{(j)}x),\; j = 1,2,..\, \bar{M}\;. \qquad (1.47)$$

A new spectral transform of Q(x), corresponding to this new spectral problem, can then be defined as

$$\overline{ST} : \{\, \bar{R}(k),\; -\infty < k < +\infty,\; \bar{p}^{(j)},\; \bar{\rho}^{(j)},\; \bar{P}^{(j)},\; j = 1,2,\ldots \bar{M}\,\} \qquad (1.48)$$

where, as before,

$$\bar{\rho}^{(j)} = \bar{c}^{(j)}\, \bar{c}^{(j)^{+}} > 0\;,\; \bar{P}^{(j)} = (\bar{\rho}^{(j)})^{-1}\, \bar{c}^{(j)^{+}}\, \bar{c}^{(j)},\; j = 1,2,\ldots\bar{M} \qquad (1.49)$$

$\bar{P}^{(j)}$ being the projection matrix corresponding to the nondegenerate eigenvalue $-\bar{p}^{(j)^2}$. However this other spectral transform (1.48) does not really add any new piece to our mathematical machinary because we now prove that

$$\text{ST of } Q(x) = \overline{ST} \text{ of } Q(x)\;. \qquad (1.50)$$

We limit our proof to the continuous part of the ST because, for the matrices Q(x) satisfying the asymptotic conditions (1.15), this result together with Eq.(1.16) implies the validity of (1.50); however the equality (1.50) holds for the larger class of matrices Q(x) considered in this context. To prove that $R(k) = \bar{R}(k)$ we note that the wronskian type expression

$$W(x,k) \equiv \bar{\Psi}(x,k)\, \Psi_x(x,k) - \bar{\Psi}_x(x,k)\, \Psi(x,k) \qquad (1.51)$$

defines a x-independent $N \times N$ matrix, namely

$$W_x(x,k) = 0\;. \qquad (1.52)$$

Let us now evaluate the matrix W by inserting in the definition (1.51) the asymptotic behaviour of the solutions $\bar{\Psi}(x,k)$ and $\Psi(x,k)$, (1.44) and (1.10) respectively and taking the limit both as $x \to -\infty$ and as $x \to +\infty$

$$W(-\infty,k) = 0 \quad , \quad W(+\infty,k) = 2ik \left[R(k) - \bar{R}(k)\right] \; ; \tag{1.53}$$

this result, together with Eq.(1.52), implies that

$$R(k) = \bar{R}(k) \tag{1.54}$$

which is just the equality (1.50) for the continuous part of the ST.

We finally notice that the hermitianity condition (1.6) implies

$$\bar{\Psi}(x,k) = \left[\Psi(x,-k^*)\right]^+ \tag{1.55}$$

and, through the asymptotic behaviour (1.10a) and (1.44a) together with the equation (1.54), also

$$R^+(k) = R(-k^*) \tag{1.56}$$

which, for real k, reduces to the equation (1.11) reported above. For completness, we write also the (generalized) unitarity equation

$$\bar{T}(-k)\; T(k) + \bar{R}(-k)\; R(k) = 1 \tag{1.57}$$

which obtains by applying the same procedure given before to the following wronskian expression

$$W_-(x,k) \equiv \bar{\Psi}(x,-k)\; \Psi_x(x,k) - \bar{\Psi}_x(x,-k)\; \Psi(x,k) \tag{1.58}$$

the details of the derivation being left to the reader as an exercise. For hermitian Q and real k Eqs.(1.55), (1.10b), (1.44b) and (1.57) imply the well-known unitarity equation

$$T^+(k)\; T(k) + R^+(k)\; R(k) = 1 \tag{1.59}$$

representing the particle flux conservation in a scattering process.

## 2. THE BASIC FORMULAE AND SNEE

The general method yielding all the results contained in these lectures is based on the identity

$$W(x_2,k) - W(x_1,k) = \int_{x_1}^{x_2} dx\, W_x(x,k) \tag{2.1}$$

where the $N \times N$ matrix $W$ has the usual wronskian-type expression

$$W(x,k) = \Psi_1(x,k)\, \Psi_{2x}(x,k) - \Psi_{1x}(x,k)\, \Psi_2(x,k)\,; \tag{2.2}$$

the equation (2.1), when specialized to the cases indicated below, produces the basic functional relationship between Q's and their spectral transforms, which are the building blocks of the present approach.

Let $\bar{\Psi}'$, respectively $\Psi$, be two matrix solutions of Eqs.(1.43) (with potential $Q'(x)$), respectively (1.9), and consider the wronskian $W$ defined by (2.2) with

$$\Psi_1(x,k) = \bar{\Psi}'(x,k)\ , \quad \Psi_2(x,k) = F(x)\Psi_x(x,k) \tag{2.3}$$

where $F(x)$ is an arbitrary (twice differentiable) matrix. Using on the r.h.s. of the identity (2.1) the differential equations (1.43) and (1.9), and integrating appropriately by parts |16|, one easily gets the generalized wronskian identity

$$\Big\{ \bar{\Psi}'(-2k^2 F - F_{xx} + FQ + Q'F)\Psi - 2\bar{\Psi}'_x F \Psi_x + \bar{\Psi}' F_x \Psi_x + \bar{\Psi}'_x F_x \Psi -$$

$$\bar{\Psi}' \Big[ \int_{x_0}^{x} dx' \, (FQ - Q'F) \Big] \Psi_x + \bar{\Psi}'_x \Big[ \int_{x_0}^{x} dx' \, (FQ - Q'F) \Big] \Psi \Big\} \Big|_{x_1}^{x_2} =$$

$$\int_{x_1}^{x_2} dx\, \bar{\Psi}' \Big\{ - F_{xxx} + 2F_x Q + 2Q'F_x - 4k^2 F_x + FQ_x + Q'_x F -$$

$$\Big[ \int_{x_0}^{x} dx' (FQ - Q'F) \Big] Q + Q' \Big[ \int_{x_0}^{x} dx' \, (FQ - Q'F) \Big] \Big\} \Psi \,. \tag{2.4}$$

Note that $x_0$ is an arbitrary parameter. We are, of course, assuming $F(x)$ and the potentials $Q(x)$ and $Q'(x)$ to be sufficiently regular to justify the various integrations by parts required for the derivation of this equation. The key idea in deriving the identity (2.4) is to use the integration by parts to end up with an integrand matrix in the r.h.s. of a sandwich type, i.e. some matrix in between $\bar{\Psi}'$ and $\Psi$, while all contributions from the x-derivative of $\bar{\Psi}'$ and $\Psi$ should come only from the ends of the integration interval.

From (2.4) now we derive three important formulae:

i) from the arbitrariness of $x_0$ we get first of all the formula

$$\left[\bar{\Psi}'_x M \Psi - \bar{\Psi}' M \Psi_x\right]\Bigg|_{x_1}^{x_2} = \int_{x_1}^{x_2} dx\, \bar{\Psi}' \{ Q'M - MQ \} \Psi \tag{2.5}$$

where M is an arbitrary x-independent matrix. Using the asymptotic expressions (1.10) and (1.44) where appropriate, one finds that the limit $x_1 \to -\infty$ of the l.h.s. of (2.5) vanishes, while its limit $x_2 \to +\infty$ is related to the reflection coefficients corresponding to Q' and Q

$$2ik\left[R'(k)\,M - M\,R(k)\right] = \int_{-\infty}^{+\infty} dx\, \bar{\Psi}'(x,k)\,\{ Q'(x)\,M - M\,Q(x) \}\,\Psi(x,k) \;. \tag{2.6}$$

This equation implies that, if a matrix M commutes with Q(x) for any x, $[M,Q(x)] = 0$, M commutes with R(k) for any k, $[M,R(k)] = 0$.

ii) Next we set F(x) = N, where N is x-independent, and, choosing $x_0 = +\infty$ and taking the limits $x_1 \to -\infty$ and $x_2 \to +\infty$, we get

$$(2ik)^2 \left[R'(k)\,N + N\,R(k)\right] = \int_{-\infty}^{+\infty} dx\, \bar{\Psi}'(x,k)\,\{ Q'_x(x)N + N\, Q_x(x) \; +$$

$$Q'(x) \int_x^{+\infty} dx' \left[Q'(x')N - NQ(x')\right] - \int_x^{+\infty} dx' \left[Q'(x')N - NQ(x')\right] Q(x) \}\,\Psi(x,k) \,. \tag{2.7}$$

iii) Finally, we consider a matrix F(x), that is arbitrary, but vanishes asymptotically. Specifically, we assume

$$F(+\infty) = F_x(\pm\infty) = F_{xx}(\pm\infty) = 0 \tag{2.8}$$

and, again with $x_0 = +\infty$ and taking in (2.4) the limits $x_1 \to -\infty$ and $x_2 \to +\infty$, we obtain

$$(2ik)^2 \int_{-\infty}^{+\infty} dx\, \bar{\Psi}'(x,k)\, F_x(x)\, \Psi(x,k) = \int_{-\infty}^{+\infty} dx\, \bar{\Psi}'(x,k)\,\{ F_{xxx}(x) - 2Q'(x)\, F_x(x) \; -$$

$$2F_x(x)\, Q(x) - Q'_x(x)F(x) - F(x)Q_x(x) - Q'(x)\left[\int_x^{+\infty} dx' \left[Q'(x')\, F(x') - F(x')Q(x')\right]\right] +$$

$$\left[\int_x^{+\infty} dx' \left[Q'(x')\, F(x') - F(x')\, Q(x')\right]\right] Q(x) \}\,\Psi(x,k) \;. \tag{2.9}$$

The content of this formula is more transparently shown by introducing the integro-differential linear operator $\Lambda$, defined by the following formulae that detail its action on a generic matrix F(x) (vanishing at + ∞):

$$\Lambda F(x) = F_{xx}(x) - 2\left[Q'(x)\,F(x) + F(x)\,Q(x)\right] + \Gamma \int_x^{+\infty} dx'\, F(x') , \tag{2.10}$$

$$\Gamma F(x) = Q_x'(x)\,F(x) + F(x)\,Q_x(x) + \int_x^{+\infty} dx' \left[Q'(x)\,Q'(x')\,F(x') - \right.$$

$$\left. - Q'(x)\,F(x')\,Q(x') - Q'(x')\,F(x')\,Q(x) + F(x')\,Q(x')\,Q(x)\right] . \tag{2.11}$$

Thus, if F(x) is an arbitrary matrix (vanishing as $|x| \to \infty$), the equation (2.9) can be rewritten in the more compact form

$$\int_{-\infty}^{+\infty} dx\, \bar{\Psi}'(x,k)\,\{\Lambda F(x)\}\,\Psi(x,k) = (2ik)^2 \int_{-\infty}^{+\infty} dx\, \bar{\Psi}'(x,k)\,F(x)\,\Psi(x,k) \tag{2.12}$$

(since here F(x) is arbitrary, no confusion should be caused by replacing the matrix $F_x$ entering in (2.9) with F(x), which is now the generic matrix in (2.12)). If we now assume that the matrix F(x) vanishes asymptotically with all its derivatives

$$\lim_{x \to \pm\infty} \left[d^n F(x)/dx^n\right] = 0 \quad , \quad n = 0,1,2,..., \tag{2.13}$$

we can let $\Lambda$ act repeatedly on F(x), obtaining

$$\int_{-\infty}^{+\infty} dx\, \bar{\Psi}'(x,k)\,\{\Lambda^n F(x)\}\,\Psi(x,k) = (2ik)^{2n} \int_{-\infty}^{+\infty} dx\, \bar{\Psi}'(x,k)F(x)\Psi(x,k), \quad n=0,1,2,...; \tag{2.14}$$

it should be emphasized that in this formula, as well as in all the similar ones given below, the operators never act on the wave functions, even though one of these is always written on the r.h.s. (and could not be written on the l.h.s., since one is generally dealing with noncommuting matrices). Finally, from (2.14) there immediately follows

$$\int_{-\infty}^{+\infty} dx\, \bar{\Psi}'(x,k)\,\{f(\Lambda)F(x)\}\,\Psi(x,k) = f(-4k^2) \int_{-\infty}^{+\infty} dx\, \bar{\Psi}'(x,k)\,F(x)\,\Psi(x,k) \tag{2.15}$$

where f(z) is an arbitrary entire function. This remarkable equation, in which F(x) is essentially arbitrary, except for the requirement (2.13), should be somehow seen as analogous to

$$\int_{-\infty}^{+\infty} dx \, \exp(-i\,k\,x) \left[ f(d/dx)\; q(x) \right] = f(ik) \int_{-\infty}^{+\infty} dx \, \exp(-i\,k\,x)\; q(x) \tag{2.16}$$

which is well-known in Fourier analysis.

The basic formulae of this approach obtain by inserting $F(x) = Q'(x)M - MQ(x)$, respectively $F(x) = \Gamma N$, in (2.15) and by using on the r.h.s. Eq.(2.6), respectively Eq.(2.7). In this manner we get

$$2i\,k\,f(-4k^2) \left[ R'(k)\,M - M\,R(k) \right] = \int_{-\infty}^{+\infty} dx\; \bar{\Psi}'(x,k)\, \{ f(\Lambda) \left[ Q'(x)M - MQ(x) \right] \}\, \Psi(x,k) \tag{2.17}$$

$$(2ik)^2 g(-4k^2) \left[ R'(k)N + NR(k) \right] = \int_{-\infty}^{+\infty} dx\; \bar{\Psi}'(x,k)\, \{ g(\Lambda)\, \Gamma\, N \}\, \Psi(x,k)\ . \tag{2.18}$$

These equations provide the main tool of our treatment. Let us re-emphasize that they have been shown to hold for arbitrary (entire) functions f and g, and for arbitrary x-independent matrices M and N. The operators $\Lambda$ and $\Gamma$ are defined by Eqs. (2.10) and (2.11); they involve the two matrices $Q(x)$ and $Q'(x)$, to which there correspond the reflection coefficients $R(k)$ and $R'(k)$.

Thus the equation (2.17) and (2.18) relate the two matrices $Q'(x)$ and $Q(x)$ to the continuous part of their ST (see definition (1.20)). Similar equations connecting $Q'(x)$ and $Q(x)$ to the discrete part of their ST (i.e. to the discrete spectrum of their corresponding Schrödinger eigenvalue problems), as well as to their corresponding transmission coefficients $T'(k)$ and $T(k)$, can also be obtained by the generalized wronskian relation technique; we do not report them here since, for our present purposes, they are not really needed (and the interested reader can find them in reference |11|).

We now proceed to discuss the first of several important applications of the basic formulae (2.17) and (2.18), namely the derivation of the class of SNEE. To this aim, we introduce a basis of the $N^2$-dimensional space of $N \times N$ matrices, namely the $N^2$ hermitian matrices $\sigma_\nu$, $\nu = 0,1,\ldots,N^2-1$, with $\sigma_0 = 1$; greek indices run over the values $0,1,2,\ldots,N^2-1$; latin indices over the values $1,2,\ldots,N^2-1$. The convention that repeated indices are summed is always understood. Furthermore, the conventional notation for commutators and anticommutators used throughout is $[A,B] = AB - BA$, $\{A,B\} = AB + BA$.

We introduce a one-parameter family of matrices $Q(x,t)$, the parameter t being the time, and we set

$$Q(x) = Q(x,t) \quad , \quad Q'(x) = Q(x,t+\Delta t) \tag{2.19}$$

that of course also imples

$$R(k) = R(k,t) \quad , \quad R'(k) = R(k,t+\Delta t) \tag{2.20}$$

and analogous relations for the parameters of the discrete spectrum and for the transmission coefficients.

Now we insert this ansatz in Eqs. (2.17) and (2.18), and investigate the limit $\Delta t \to 0$, obtaining thereby a set of relations that provide the basis for establishing a class of nonlinear evolution equations solvable by the ST associated to the matrix Schrödinger problem.

Let us begin by writing down the operators $\Lambda$ and $\Gamma$, defined by Eqs.(2.10) and (2.11), respectively, in the limit $Q'(x) = Q(x)$. We call these operators L, respectively G; they are defined by detailing their action on a generic matrix F(x) (vanishing at infinity):

$$LF(x) = F_{xx}(x) - 2\{Q(x,t),\, F(x)\} + G \int_x^{+\infty} dx'\; F(x') \tag{2.21}$$

$$GF(x) = \{Q_x(x,t),\, F(x)\} + \left[Q(x,t),\, \int_x^{+\infty} dx'\; [Q(x',t),\, F(x')]\right] . \tag{2.22}$$

Then set $M = \sigma_n$ in Eq.(2.17), and write $f_n$ in place of f; for $\Delta t = 0$, this yields

$$2ikf_n(-4k^2)\left[\sigma_n, R(k,t)\right] = \int_{-\infty}^{+\infty} dx\; \bar{\Psi}(x,k,t)\,\{\, f_n(L)\left[\sigma_n, Q(x,t)\right]\}\,\Psi(x,k,t)\,. \tag{2.23}$$

Had we set instead $M = \sigma_0$, we would have obtained for $\Delta t = 0$ the trivial identity $0 = 0$; therefore, in order to obtain a nontrivial equation, we set $M = \sigma_0/\Delta t$ in Eq.(2.17), and take the limit $\Delta t \to 0$. This yields

$$2\,i\,k\,f(-4k^2)\,R_t(k,t) = \int_{-\infty}^{+\infty} dx\; \bar{\Psi}(x,k,t)\,\{\, f(L)\, Q_t(x,t)\,\}\,\Psi(x,k,t)\,. \tag{2.24}$$

Set finally, in Eq.(2.18), $N = \sigma_\mu$ and write $g_\mu$ in place of g. For $\Delta t = 0$, this yields

$$(2ik)^2 \, g_\mu(-4k^2) \, \{\sigma_\mu, R(k,t)\} = \int_{-\infty}^{+\infty} dx \, \bar{\Psi}(x,k,t) \, \{g_\mu(L) G \, \sigma_\mu\} \Psi(x,k,t) \, . \qquad (2.25)$$

Note that the indices n and $\mu$ in Eq.(2.23) and in Eq.(2.25), respectively, may or may not be understood as summed upon, since the functions $f_n$ and $g_n$, due to their arbitrariness, might all but one chosen to vanish. Furthermore, the arbitrariness of f, $f_n$ and $g_\mu$ in the 3 equations written above implies that these functions could also parametrically depend on t; they must of course be independent of x.

From Eqs.(2.23), (2.24) and (2.25), the following result is immediately implied: if the matrix Q(x,t) satisfies the nonlinear equation

$$f(L,t) \, Q_t = \alpha_n(L,t) \left[\sigma_n, Q\right] + \beta_\nu(L,t) \, G \, \sigma_\nu \qquad (2.26)$$

the corresponding reflection coefficient R(k,t) satisfies the linear ordinary differential equation

$$f(-4k^2,t) \, R_t = \alpha_n(-4k^2,t) \left[\sigma_n, R\right] + 2ik \, \beta_\nu(-4k^2,t) \, \{\sigma_\nu, R\} \, . \qquad (2.27)$$

In the following, we restrict, for simplicity, our attention to the case $f = 1$, and moreover we assume the (entire but otherwise arbitrary) functions $\alpha_n$ and $\beta_\nu$ to be independent of t. Thus we focus attention on the SNEE

$$Q_t(x,t) = \alpha_n(L) \left[\sigma_n, Q(x,t)\right] + \beta_\nu(L) G \, \sigma_\nu \, . \qquad (2.28)$$

The nonlinearity of this equation originates from the dependence of the operator L on Q, as explicitely shown by its definition Eqs.(2.21) and (2.22). In order to solve the initial value problem for the nonlinear evolution equation (2.28), according to the solvability diagram we have previously discussed, we need to be able to solve the corresponding initial value problem for the ST of Q(x,t). For the continuous part of the ST we have the following ordinary linear matrix differential equation

$$R_t(k,t) = \alpha_n(-4k^2) \left[\sigma_n, R(k,t)\right] + 2ik \, \beta_\nu(-4k^2) \, \{\sigma_\nu, R(k,t)\} \qquad (2.29)$$

which obtains by specializing Eq.(2.27) to the present subclass of SNEE.

As for the time evolution of the discrete part of the ST of Q(x,t), one might apply the generalized wronskian method as it has been done for the reflection coefficient; however, a more straightforward procedure, yielding the same evolution equations, even though it applies only in the case of matrices Q that vanish asymptotically faster than exponentially, takes advantage of the relationship between the discrete spectrum parameters $p^{(j)}$ and $c^{(j)}$ and the locations and residues of the poles of the reflection coefficient (see Eq.(1-16)).

For the evolution of $p^{(j)}(t)$, being $k = ip^{(j)}(t)$ a simple pole of $R(k,t)$, one immediately obtains from (2.29)

$$p_t^{(j)}(t) = 0 \quad , \quad j = 1,1,\ldots M \tag{2.30}$$

To write down the evolution equation of the remaining part of the ST, it is convenient to introduce the matrices

$$C^{(j)}(t) = c^{(j)}\left[c^{(j)}(t)\right]^+ = \rho^{(j)}(t)\; P^{(j)}(t) \quad , \quad j = 1,2,\ldots M \tag{2.31}$$

since they are simply related to the residue of $R(k,t)$ at $k = ip^{(j)}(t)$; then for these matrices, Eq.(2.29) implies

$$C_t^{(j)}(t) = \alpha_n(4p^{(j)^2})\left[\sigma_n, C^{(j)}(t)\right] - 2p^{(j)}\; \beta_\nu(4p^{(j)^2})\;\{\sigma_\nu,\; C^{(j)}(t)\} \tag{2.32}$$

The constancy of the eigenvalues $-p^{(j)^2}$, i.e. their time independence, as implied by Eq.(2.30), shows that the flow (2.28) is isospectral, a characteristic feature of many important evolution equations discovered so far |17|.

We are now in the position to prove the solvability of the nonlinear equation (2.28). In fact, given the initial value |18|

$$Q_0(x) = Q(x,0) \tag{2.33}$$

we solve the direct problem, which is linear, and compute the ST of $Q_0(x)$

$$\text{ST of } Q(x,0) = \{R_0(k),\; p_0^{(j)},\; \rho_0^{(j)},\; P_0^{(j)},\;\; j = 1,2,\ldots M\} \tag{2.34}$$

then, we integrate the linear differential equations (2.29), (2.30) and (2.32), with the initial condition (2.34), and obtain the explicit solutions

$$R(k,t) = \exp\left[4ik\;\beta_0(-4k^2)t\right]\;\exp\{t\left[\alpha_n(-4k^2) + 2ik\;\beta_n(-4k^2)\right]\sigma_n\}\cdot$$
$$R_0(k)\;\exp\{\,t\left[-\alpha_n(-4k^2) + 2ik\;\beta_n(-4k^2)\right]\sigma_n\}\;, \tag{2.35}$$

$$p^{(j)}(t) = p_0^{(j)} \tag{2.36}$$

$$C^{(j)}(t) = \rho^{(j)}(t)\, P^{(j)}(t) = \rho_o^{(j)} \exp\left[-4p^{(j)}\, \beta_o(4p^{(j)2})t\right] \cdot$$

$$\exp\{ t \left[\alpha_n(4p^{(j)2}) - 2p^{(j)}\, \beta_n(4p^{(j)2})\right] \sigma_n \}\, P_o^{(j)} \cdot$$

$$\exp\{ t \left[-\alpha_n(4p^{(j)2}) - 2p^{(j)}\, \beta_n(4p^{(j)2})\right] \sigma_n \}, \tag{2.37}$$

and therefore the ST at the time t. Finally, by insertion of (2.35), (2.36) and (2.37) into Eq. (1.34), we solve the linear Gel'fand-Levit a n-Marchenko integral equation (1.35) of the inverse problem and obtain through Eq.(1.36), the solution Q(x,t) of the nonlinear evolution equation (2.28) at the time t.

The great simplification obtained by transforming the nonlinear evolution equation (2.28) into the corresponding linear evolution equations for the ST (2.29), (2.30) and (2.32) is the first good example of the following "golden rule": to investigate a SNEE of the class (2.28) (or, for this matter, of the class (2.26))and the properties of its solutions, go from x-space to k-space where the problem is linear. Of course, the main difficulty in solving the Cauchy problem is found in performing the transformation from the x-space to the k-space and from the k-space back to the x-space, these transformations being defined through the solution of the direct and inverse problems associated to the matrix Schroedinger operator. We emphasize that, in addition to being linear, the evolution equations in the k-space have also the virtue of not coupling to each other the different components of the ST; thus, the time evolution of the parameters p and C (or $\rho$ and P, see (2.31)) characterizing each discrete eigenvalue is quite independent of the possible presence of the continuous component, or of other discrete eigenvalues. As it will be clear in the next section, it is this fact, coupled with the solvability through the ST method, that originates the remarkable behaviour of "solitons", in particular their stability. It also motivates the interest in the single-soliton solution, since this, not only provides a remarkable special solution of the class of SNEEs being investigated, but indeed constitutes a component (that may, or may not, show up asymptotically; see below) of a broad class of solutions.

We end this section writing down in our formalism few examples of nonlinear evolution equations which are already well-known |5|:

i) the Korteweg-de Vries equation

$$u_t(x,t) = u_{xxx}(x,t) - 6u(x,t)\, u_x(x,t) \tag{2.38}$$

obtains by setting N = 1, to say the matrix Q(x,t) = u(x,t) reduces to a scalar field, and $\beta_o(z) = z/2$.

ii) the modified Korteweg-de Vries equation

$$q_t(x,t) = q_{xxx}(x,t) + 6q^2(x,t)\, q_x(x,t) \tag{2.39}$$

obtains with $N = 2$, by writing the $2 \times 2$ matrix $Q(x,t)$ in terms of the scalar function $q(x,t)$ according to the following ansatz

$$Q(x,t) = \begin{pmatrix} -q^2(x,t) & q_x(x,t) \\ -q_x(x,t) & -q^2(x,t) \end{pmatrix}, \tag{2.40}$$

and by choosing $\alpha_n(z) = \beta_n(z) = 0 \quad (n = 1,2,3)$ and $\beta_o(z) = z/2$.

iii) the nonlinear Schrödinger equation

$$i\, \psi_t(x,t) = -\psi_{xx}(x,t) - 2|\psi(x,t)|^2\, \psi(x,t) \tag{2.41}$$

obtains with $N = 2$, $\beta_\nu(z) = 0$, $\alpha_1(z) = \alpha_2(z) = 0$, $\alpha_3(z) = iz/2$ and relating the $2 \times 2$ matrix $Q(x,t)$ to the function $\psi(x,t)$ as follows

$$Q(x,t) = \begin{pmatrix} -|\psi(x,t)|^2 & \psi_x(x,t) \\ -\psi_x^*(x,t) & -|\psi(x,t)|^2 \end{pmatrix}. \tag{2.42}$$

iv) the sine-Gordon equation

$$\phi_{xt}(x,t) = \sin\left[\phi(x,t)\right] \tag{2.43}$$

obtains with $N = 2$, $\alpha_n(z) = \beta_n(z) = 0 \quad (n = 1,2,3)$, $\beta_o(z) = (2z)^{-1}$ and

$$Q(x,t) = -(1/4) \begin{pmatrix} \phi_x^2(x,t) & 2\phi_{xx}(x,t) \\ -2\phi_{xx}(x,t) & \phi_x^2(x,t) \end{pmatrix}. \tag{2.44}$$

v) the boomeron equation |19|

$$U_t(x,t) = \vec{b} \cdot \vec{V}_x(x,t) \tag{2.45a}$$

$$V_{xt}(x,t) = U_{xx}(x,t)\vec{b} + \vec{a} \wedge \vec{V}_x(x,t) - 2\vec{V}_x(x,t) \wedge \left[\vec{V}(x,t) \wedge \vec{b}\right] \tag{2.45b}$$

is a system of coupled nonlinear equations for the scalar field $U(x,t)$ and the three-dimensional vector field $\vec{V}(x,t)$, and obtains with $N = 2$, $\beta_0(z) = 0$, $\alpha_n(z) = a_n/2i$ and $\beta_n(z) = b_n/2$ $(n = 1,2,3)$ being constants and defining the two three-dimensional vectors $\vec{a}$ and $\vec{b}$, and

$$\int_x^{+\infty} dx'\ Q(x',t) = U(x,t) + \vec{\sigma}\,.\,\vec{V}(x,t) \tag{2.46}$$

where $\vec{\sigma} \equiv (\sigma_1, \sigma_2, \sigma_3)$ are the three Pauli matrices.

Nothe that a rigorous treatment of the important equations ii), iii) and iv) requires an extension of the present formalism to non hermitian |20| matrices $Q(x,t)$. Although this can be easily done, in the following we will consider only cases where the hermitianity condition (1.6) holds. Therefore, since the basis matrices $\sigma_\nu$ are hermitian, sufficient conditions to guarantee that $Q(x,t)$ remain hermitian throughout its time evolution if it is hermitian at the initial time $t_0$, are that the functions $\beta_\nu$ and $\alpha_n$ in Eq. (2.28) be real and imaginary respectively

$$\alpha_n = -\alpha_n^* \;,\quad \beta_\nu = \beta_\nu^* \;. \tag{2.47}$$

In the next section the motion of the solitons, particularly the boomeron equation, will be investigated, and their non vanishing acceleration will be exhibited.

## 3. SOLITONS

To set up the proper language to describe the nonlinear phenomena modeled by the nonlinear evolution equations solvable by the ST method, the continuous part of the ST will be hereafter referred to as the background component of the ST. The reason is that a solution of a SNEE, whose ST has no discrete part, evolves in time very similarly to a wave-packet undergoing the well-known dispersion phenomenon. On the other hand, we will refer to the discrete part of the ST as to the soliton component; in fact, solutions with a vanishing reflection coefficient, i.e. with

$ST = \{R(k) = 0,\ p^{(j)},\ \rho^{(j)},\ P^{(j)},\ j = 1,2,\dots, M\}$,

behave in a quite different way, since a sort of balance between dispersive and nonlinear effects give them stability and localizability properties which show up in interesting and beautiful physical phenomena. An enthusiastic description of a phenomenon of this sort in hydrodynamics, has been given by J. Scott-Russel |21| in the far 1844, this being an early introduction of solitons to observational science, well before the rather recent theoretical description. An important rôle in discovering solitons and understanding their properties has been played by computer experiments, among which the most crucial have been those produced in 1965 by N.J. Zabusky and M. Kruskal |22,23|, who also introduced the name "soliton". Since then the number of papers devoted to solitons has been growing up to now in an explosive way, and an up-do-date review of soliton phenomenology in some branches of physics has been presented by R.K. Bullough |3|.

Here we focus our discussion only on those solutions of the nonlinear evolution equation (2.28) whose ST contains only the soliton component, namely with a vanishing background component.

Let us start with the more familiar 1-soliton solution in the simpler case $N=1$, so that eq.(2.28) reduces to the scalar field equation

$$Q_t(x,t) = 2\beta_0(L)\, Q_x(x,t) \tag{3.1}$$

the easiest way to construct this solution is to consider its spectral transform at a given time t, to say, $M=1$, $R(k,t) = 0$, $\rho^{(1)}(t) = \rho(t)$, $p^{(1)}(t) = p$, and apply the techniques described in the previous sections. The corresponding inverse spectral transform was already obtained solving the Marchenko equation, and given by the expression (1.37) which in the present case reads

$$Q(x,t) = -A/\cos h^2 \{ \left[x - \xi(t)\right]/\lambda \} \tag{3.2}$$

where we have taken into account the expressions (1.38) of the amplitude A and width $\lambda$ of the soliton in terms of the spectral parameter p, and the fact that this parameter p is constant because of the isospectral evolution expressed by the equation (2.30). Therefore the t-dependence of the solution $Q(x,t)$ enters only through the function

$$\xi(t) = (1/2p)\ \ln \left[\rho(t)/2p\right] \tag{3.3}$$

and is immediately obtained by specializing the general solution (2.37) to the scalar case, namely

$$\xi(t) = \xi_0 - 2\beta_0(4p^2)\ t \ . \tag{3.4}$$

With these findings in our hands, we should draw our attention on the following facts: i) the amplitude A and the width $\lambda$ of the soliton do not depend either on time or on the particular SNEE belonging to the class (3.1), and are related to each other

$$A = 2/\lambda^2 \quad ; \tag{3.5}$$

ii) the function $\xi(t)$ should be interpreted as the position of the soliton at the time t and the expression of the spectral parameter $\rho(t)$ in terms of the soliton motion is

$$\rho(t) - 2p \exp\left[2p\xi(t)\right] \quad ; \tag{3.6}$$

iii) the soliton does not accelerate and behaves as a free particle moving with the constant speed

$$v = -2\,\beta_0(4p^2) \quad . \tag{3.7}$$

This particle-like behaviour has certainly motivated the interest of many physicists which consider this mathematical framework, namely nonlinear theory, as a first step towards an elementary particle theory which avoids the unpleasant point-like model; however, this is only but one of many others, and important, applications of the concept of soliton to physics.

In order to investigate which kind of soliton dynamics is implied by these nonlinear evolution equations, we have to investigate the large $|t|$ evolution of a many-soliton solution. After the numerical experiments showing a two-soliton collision carried out by N.J. Zabusky |22|, the first proof that two colliding solitons eventually reappear unchanged is due to P.D. Lax |6| (they both investigated the KdV equation). Let us consider then a solution containing only M solitons, its spectral transform being $\{R(k,t) = 0,\ \rho^{(j)}(t),\ p^{(j)},\ j = 1,2\ldots,M\}$; although at finite time the nonlinear effects may prevent us to recognize the individual solitons, they will ultimately show up well separated from each other because of their different velocities $v^{(j)} = -2\,\beta_0(4p^{(j)2})$ (see Eq.(3.7)). Indeed, the well-known stability of the soliton is nicely expressed by the asymptotic behaviour of the M-soliton solution for large values of $|t|$, which shows M 1-soliton bumps with the usual shape

$$Q_\pm^{(j)}(x,t) = -A^{(j)}/\cosh^2\{\left[x - \xi_\pm^{(j)}(t)\right]/\lambda^{(j)}\},\ \text{as } t \to \pm\infty \quad . \tag{3.8}$$

The effect of their nonlinear interaction is accounted for by the following simple rule

$$\xi_+^{(j)}(t) = \xi_-^{(j)}(t) + d^{(j)},\quad d\xi_\pm^{(j)}(t)/dt = v^{(j)} = -2\beta_0(4p^{(j)2}) \tag{3.9}$$

the amplitudes and widths having the usual expression $A^{(j)} = 2p^{(j)2}$ and $\lambda^{(j)} = 1/p^{(j)}$. Therefore, while the amplitudes, widths and speeds have not been changed by the interaction, the only dynamical effect is the displacement $d^{(j)}$ of the j-th soliton position with respect to its initial free motion. The expression of the displacement $d^{(j)}$ in terms of the soliton widths $\lambda^{(j)}$'s, obtained independently |24| (for the KdV equation) by V.E. Zakharov (1971), M. Toda and M. Wadati (1972) and S. Tanaka (1972), shows that these displacements are given by the sum of M-1 two-soliton contributions, and this proves that solitons interact only via two-body forces. Furthermore the relationship satisfied by the displacements $d^{(j)}$, namely $\sum_{j=1}^{M} (d^{(j)}/\lambda^{(j)}) = 0$ is naturally understood as the free motion of the center of mass of the M-soliton system. The qualitative picture emerging from these results is that of an isolated system of M classical particles. The attractive character of the two-soliton force is easily shown in the case of a two-soliton collision; if $v^{(1)} > v^{(2)}$, the soliton 1 is advanced while the other one is delayed, their corresponding shifts being

$$d^{(1)} = \lambda^{(1)} \ln|(\lambda^{(1)} + \lambda^{(2)})/(\lambda^{(1)} - \lambda^{(2)})| > 0,\quad d^{(2)} = \lambda^{(2)} \ln|(\lambda^{(1)} - \lambda^{(2)})/(\lambda^{(1)} + \lambda^{(2)})| < 0 \qquad (3.10)$$

With the aim of finding solitons with a richer dynamics (for instance, solitons with more degrees of freedom interacting via attractive or repulsive forces), let us look at a system of coupled nonlinear evolution equations for a multi-component field, namely for a matrix Q(x,t). We have already shown in the previous section that important nonlinear equations such as the sine-Gordon, nonlinear Schrödinger and modified KdV equations can be derived, as special cases of non hermitian 2 x 2 matrices, from the class of SNEE (2.28). However also in these cases the interacting solitons experience in the collision only a displacement of their position; interesting features are nevertheless exhibited by the sine-Gordon equation which, among other virtues (for instance, its relativistic invariance), has solutions describing two kinks (solitons) colliding via repulsive interaction as well as a kink-antikink bound state |25| ("breather" or "bion"). In order to explore the general properties of the soliton solutions of the equation (2.28), we consider first the case of N x N hermitian matrix fields.

Let us focus first on the 1-soliton solution. This is easily obtained inserting in the Marchenko equation (1.35) its spectral transform $\{R(k,t) = 0, p, \rho(t), P(t)\}$. This result reads

$$Q(x,t) = - \{A/\cosh^2 \{ [x - \xi(t)]/\lambda \} \} P(t) . \qquad (3.11)$$

Again the constant (scalar) A and the width $\lambda$ are simply related to the spectral parameters, $A = 2p^2$ and $\lambda = p^{-1}$. The position of the soliton at time t, $\xi(t)$, is related to spectral parameter $\rho(t)$ by the formula (3.6).

The novelty of this solution is due to the remarkable fact that $\xi(t)$ is generally not linear in t, i.e. this soliton generally moves with a speed which varies in time. The mechanism which is responsible for this interesting feature originates from the time dependence of the projection matrix P(t), and its relation to the soliton speed (see below). This implies that one of the $N^2$ components of our field solution, say a matrix element $Q_{ij}(x,t)$, although it has the usual $(\cosh)^{-2}$ shape in the x variable, generally has the t-dependent amplitude $A\,P_{ij}(t)$ which makes them look different from the usual solitons. Our soliton therefore is characterized by its position $\xi(t)$ and by the one-dimensional projection matrix P(t) that shall be occasionally referred to as the polarization of the soliton.

The effect of the polarization time-dependence on the motion of the soliton is better investigated considering the time evolution of the corresponding spectral transform; if c(t) is the N-dimensional vector such that $c(t)c^{+}(t) = \rho(t)\,P(t)$ (see Eq.(2.31)), then Eq.(2.32) and Eq.(3.6) imply the following equations of motion for the soliton position and polarization

$$\xi_t(t) = -2(c(t), \beta c(t))/||c(t)||^2 - 2\beta_o(4p^2) \tag{3.12}$$

$$P_t(t) = [(\alpha - 2p\beta), P(t)] - 4p\,P(t)\,\beta[1 - P(t)] \tag{3.13}$$

where we have set

$$\alpha \equiv \alpha_n(4p^2)\,\sigma_n \quad , \quad \beta \equiv \beta_n(4p^2)\,\sigma_n \,. \tag{3.14}$$

The SNEE's discussed by Wadati and Kamijo |12| imply $\alpha = 0$ and require that $\beta$ commutes with P(t); therefore in this case Eqs.(3.13) and (3.12) describe the usual soliton with P(t) = P(0) and $\xi_t(t) = \xi_t(0)$. A soliton with constant speed can be a solution also of our more general equations (3.12) and (3.13) if its initial polarization $P_o \equiv P(0)$ is a solution of the matrix equation

$$\left[(\alpha - 2p\beta), P_o\right] - 4p\,P_o\,\beta(1 - P_o) = 0 \,. \tag{3.15}$$

If this is the case, $P(t) = P_o$ is the solution and the velocity of the soliton turns out to be constant. Indeed, if the unit vector u(t) is introduced

$$u(t) \equiv c(t)/||c(t)|| \quad , \; P(t) = u(t)\,u^{+}(t) \tag{3.16}$$

then it is easily found that the solutions of the matrix equation (3.15), wich are projectors on one dimensional subspaces, are given by the following eigenvalue equation

$$P_o = u_o u_o^+ \quad , \quad (\alpha - 2p\,\beta)\,u_o = \mu u_o \tag{3.17}$$

This implies that, if $[\alpha,\beta] = 0$, being $\beta = \beta^+$ and $\alpha = -\alpha^+$ (see (2.47)and (3.14)), the matrix $\alpha - 2p\beta$ is normal and its N eigenvectors $u_k$, $k = 1,2,..,N$, define N 1-soliton solutions with constant speed $v_k = -2(u_k, \beta u_k) - 2\beta_o(4p^2)$ and polarization $P_k = u_k u_k^+$. Of course these solutions are very special and, as it will be clear in the following, very unstable with respect to small changes of the initial value of the polarization matrix. In order to discuss the general behaviour of our solitons, we give the explicit solution of the equations of the motion (3.12) and (3.13)

$$\xi(t) = \xi_o - 2\beta_o(4p^2)t + (1/2p)\ \ln\left[(u_o,E(t)u_o)\right] \tag{3.18}$$

with

$$E(t) \equiv \{\exp\left[t(\alpha - 2p\beta)\right]\}^+ \quad \{\exp\left[t(\alpha - 2p\beta)\right]\}\ . \tag{3.19}$$

The corresponding formula for P(t) is

$$P(t) = \left[(u_o,E(t)u_o)\right]^{-1} \exp\left[t(\alpha - 2p\beta)\right]\ P_o\ \exp\left[t(\alpha - 2p\beta)^+\right] \tag{3.20}$$

where, of course, $u_o = u(0)$ and $P_o = u_o u_o^+ = P(0)$. To discuss the behaviour of the soliton it is convenient to introduce the spectral decomposition of the hermitian matrix E(t):

$$E(t) = \sum_{k=1}^{N} \exp\left[2p\ \zeta_k(t)\right]\ E_k(t) \tag{3.21}$$

where, of course,

$$E_k(t)\ E_l(t) = \delta_{kl} E_k(t)\quad . \tag{3.22}$$

Note that the quantities $\zeta_k(t)$, as well as the projection operators $E_k(t)$, depend on the functions $\alpha_n$ and $\beta_n$ that characterize the structure of the SNEE under consideration; they depend on the initial conditions only through the value of p. The time evolution of $\xi(t)$ and P(t) depends moreover on the initial vector $u_o \equiv u(0)$, through the quantities

$$e_k(t) \equiv (u_o,\ E_k(t)\ u_o)\quad . \tag{3.23}$$

The explicit formulae are

$$\xi(t) = \xi_o - 2\beta_o(4p^2)t + (1/2p)\ \ln\{\sum_{k=1}^{N} e_k(t)\exp\left[2p\ \zeta_k(t)\right]\}\ , \tag{3.24}$$

$$P(t) = \{ \sum_{k=1}^{N} e_k(t)\exp\left[2p\zeta_k(t)\right]\}^{-1}.\exp\left[t(\alpha - 2p\beta)\right] P_o.\exp\left[t(\alpha - 2p\beta)^+\right] . \qquad (3.25)$$

In order to understand the effects on the soliton motion due to the matrices $\alpha$ and $\beta$, let us consider first the special case $\beta = 0$. Then $E(t) = 1$, $\xi(t) = \xi_o - 2p\beta_o(4p^2)t$; the soliton moves with constant velocity, while its polarization matrix P(t) evolves according to the unitary transformation $\exp(t\alpha)$, namely its corresponding unit vector u(t) undergoes a precessional motion (this because $\alpha$ is antihermitian). Since the t-dependence due to $\beta_o$ is relatively trivial, in the following we set for simplicity $\beta_o \doteq 0$. If, on the contrary, the matrix $\beta$ is not vanishing, the soliton experiences on acceleration and its motion may be very complicate; however it is rather easy to discuss its asymptotic motion, i.e. for $|t| \to \infty$, if we assume that

$$\left[\alpha,\beta\right] = 0 \ . \qquad (3.26)$$

In this case the matrix E(t) does not depend on the matrix $\alpha$ and therefore, as it is implied by Eq.(3.18), the matrix $\alpha$ has no effect on the position of the soliton at any time; the matrix $\alpha$ contributes only to the time evolution of the projector P(t) by superimposing to the motion due to the matrix $\beta$ the "precession effect" given by the unitary transformation $\exp(t\alpha)$. Furthermore this effect due to the matrix $\alpha$ on the polarization P(t) disappears as $|t| \to \infty$ (see below) so that the asymptotic motion of the soliton depends only on the matrix $\beta$. Indeed, in this case,

$$E_k(t) = E_k \quad , \quad \zeta_k(t) = -2t\beta^{(k)} \quad , \ k = 1,2,..,N \qquad (3.27)$$

where the $\beta^{(k)}$'s, which are the eigenvalues of the matrix $\beta$, are ordered according to the prescription $\beta^{(1)} < \beta^{(2)} < \ldots < \beta^{(N)}$ (assuming no degeneracy), and the projection operators $E_k$ are those projecting on the eigenvectors of the matrix $\beta$. The quantities $e_k(t)$ are t-independent, and, if we exclude the special solutions mentioned above which correspond to the initial condition $e_k(0) = \delta_{k1}$, the soliton moves at large $|t|$ as a free particle

$$\xi(t) = \xi_o + (1/2p)\ln\left[e_m(0)\right] - 2t\beta^{(m)} + O\{\exp\left[-4t(\beta^{(m+1)} - \beta^{(m)})\right]\} \qquad (3.28a)$$

as $t \to +\infty$

$$\xi(t) = \xi_o + (1/2p)\ln\left[e_M(0)\right] - 2t\beta^{(M)} + O\{\exp\left[4t(\beta^{(M)} - \beta^{(M-1)})\right]\} \qquad (3.28b)$$

as $t \to -\infty$

where the indeces m and M are such that

$$m < M \ , \quad e_k(0) = 0 \ \text{ if } \ k < m \ , \ k > M \qquad (3.29)$$

and in general $m=1$ and $M=N$. If $\beta^{(M)}>0>\beta^{(m)}$, the soliton moves towards the right as $t\to+\infty$, and it recedes also to the right as $t\to-\infty$. This behaviour justifies the introduction of the term "boomeron". It is also easy to prove that in this case, the polarization matrix has the asymptotic value

$$\lim_{t\to+\infty} P(t) = E_m \quad , \quad \lim_{t\to-\infty} P(t) = E_M \tag{3.30}$$

where again the indeces m and M are defined by Eq.(3.29).

In the general case, the behaviour of the soliton may be completely different. However, it can be proved that, if the soliton escapes to infinity, it can do so at most with a speed that is asymptotically constant. If it is not so, and that this may be the case, it is demonstrated by the example reported below, the soliton instead oscillates indefinitely, and this one we will refer to as a "trappon".

The straightest way to display explicitely the "boomeron" and "trappon" behaviour is to consider the simplest novel SNEE of the class, namely the boomeron equation (2.45). Clearly the quantities $U(x,t)$ and $\vec{V}(x,t)$ satisfy the asymptotic conditions

$$U(+\infty,t) = U_x(\pm\infty,t) = 0 \quad , \quad \vec{V}(+\infty,t) = \vec{V}_x(\pm\infty,t) = 0 \ . \tag{3.31}$$

In view of the remarkable features of the soliton solutions described below, the discovery of a physical phenomenon modeled by the boomeron equation should be considered a very interesting goal.

In the present vector notation, the polarization of the soliton will be represented by the unit vector $\hat{n}(t)$ defined by the following equation

$$P(t) = (1/2)(1 + \hat{n}(t).\vec{\sigma}) \ . \tag{3.32}$$

Then the 1-soliton solution with amplitude A and width $\lambda$ (see Eq.(3.11)) reads

$$U(x,t) = -(2A)^{1/2}.\{1+\exp\{2\left[x-\xi(t)\right]/\lambda\}\}^{-1} \quad , \quad \vec{V}(x,t) = U(x,t)\ \hat{n}(t) \ . \tag{3.33}$$

The position and polarization evolve in time according to the equations |26|

$$\hat{n}_t(t) = \vec{a}\wedge\hat{n}(t) + (2/\lambda)\ \hat{n}(t)\wedge\left[\hat{n}(t)\wedge\vec{b}\right] \tag{3.34a}$$

$$\xi_t(t) = -\vec{b}.\hat{n}(t) \ . \tag{3.34b}$$

The explicit solution of these equations is

$$\xi(t) = \xi_0 + (\lambda/2)\ \ln\left[n_+E_+(t) + n_-E_-(t) + \bar{s}S(t) + \bar{c}C(t)\right] \tag{3.35}$$

$$\hat{n}(t) = \left[\vec{n}_+E_+(t) + \vec{n}_-E_-(t) + \vec{s}S(t) + \vec{c}C(t)\right] / \left[n_+E_+(t) + n_-E_-(t) + \bar{s}S(t) + \bar{c}C(t)\right] \tag{3.36}$$

$$E_\pm(t) = \exp(\pm a\,\nu_-\,t), \quad S(t) = \sin(a\,\nu_+\,t), \quad C(t) = \cos(a\,\nu_+\,t) \tag{3.37}$$

$$\nu_\pm = z^{-1}\{\left[z^2\cos^2\theta + (z^2-1)^2/4\right]^{1/2} \pm (z^2-1)/2\}^{1/2}, \quad z = \lambda a/(2b),$$
$$\cos\theta = \hat{a}.\hat{b} \tag{3.38}$$

$$\vec{n}_\pm = (1/2)\{(\nu_+^2 - 1)\hat{n}_0 + \left[(\hat{a}.\hat{n}_0) \mp \eta\nu_+\right]\hat{a} + z^{-2}\left[(\hat{b}.\hat{n}_0) \mp z\nu_-\right]\hat{b} \pm$$
$$\nu_-(\hat{a}\wedge\hat{n}_0) \mp \eta z^{-1}\nu_+(\hat{b}\wedge\hat{n}_0) - z^{-1}(\hat{a}\wedge\hat{b})\}/(\nu_+^2 + \nu_-^2), \quad \eta = \text{sign}\,(\hat{a}.\hat{b}) \tag{3.39}$$

$$\vec{s} = \eta\left[\nu_-\hat{a} - \eta z^{-1}\nu_+\hat{b} + \eta\nu_+(\hat{a}\wedge\hat{n}_0) + z^{-1}\nu_-(\hat{b}\wedge\hat{n}_0)\right]/(\nu_+^2 + \nu_-^2),$$
$$\vec{c} = \vec{n}_0 - \vec{n}_+ - \vec{n}_- \tag{3.40}$$

$$\bar{s} = \left[\eta\nu_-(\hat{a}.\hat{n}_0) - z^{-1}\nu_+(\hat{b}.\hat{n}_0)\right]/(\nu_+^2 + \nu_-^2), \quad \bar{c} = 1 - n_+ - n_- . \tag{3.41}$$

In the simpler case $\hat{a}\wedge\hat{b} = 0$ (we always use the notation $\vec{v} = v\,\hat{v}$) we find that if the initial polarization $\hat{n}(0) = \hat{n}_0$ coincides with one of the two unit vectors $\hat{n}_\pm = \mp\hat{b}$, then we have the two special solutions: $\hat{n}(t) = -\hat{b}$, $\xi(t) = \xi_0 + bt$ and $\hat{n}(t) = \hat{b}$, $\xi(t) = \xi_0 - bt$. If $\hat{n}(0) \neq \pm b$, $\hat{n}(t) \neq \hat{b}$ and $\hat{n}(t) \neq -\hat{b}$ for all finite t, while for the asymptotic motion we find $\hat{n}(\pm\infty) = \hat{n}_\pm = \mp\hat{b}$, $\xi_t(\pm\infty) = \pm b$; this shows a boomeron coming from infinity with the velocity -b and going back to infinity with opposite velocity. This motion is well characterized by the following theorem: the boomeron coordinate $\xi(t)$ coincides with that of a (nonrelativistic) particle of unit mass, with initial position $\xi(0) = \xi_0$ and initial speed $\xi_t(0) = -\vec{b}.\hat{n}_0$, moving in the external potential $\Phi(\xi - \xi_0)$, with

$$\Phi(x) = (1/2)\,b^2\left[1 - (\hat{b}.\hat{n}_0)^2\right]\exp(-4x/\lambda) . \tag{3.42}$$

In the general case, excluding the special one $(\hat{a}.\hat{b}) = 0$ that is discussed separately below, if the initial polarization $\hat{n}_0$ coincides with one of the two unit vectors

$$\hat{n}_\pm = (\mp\eta\,\nu_+\hat{a} \mp z^{-1}\nu_-\hat{b} - z^{-1}\hat{a}\wedge\hat{b})/(1 + \nu_-^2) \tag{3.43}$$

we have the two special solutions

$$\hat{n}(t) = \hat{n}_\pm, \quad \xi(t) = \xi_0 \pm vt \tag{3.44}$$

describing a soliton moving with a constant speed of modulus

$$v = bz\nu_- = b\{\left[z^2\cos^2\theta + (1/4)(1-z^2)^2\right]^{1/2} + (1/2)(1-z^2)\}^{1/2} . \qquad (3.45)$$

Otherwise both $\hat{n}(t)$ and $\xi_t(t)$ vary with time; at no finite time $\hat{n}(t)$ coincides with $\hat{n}_+$ or $\hat{n}_-$, while asymptotically

$$\hat{n}(\pm\infty) = \hat{n}_\pm , \quad \xi_t(\pm\infty) = \pm v . \qquad (3.46)$$

Note that this implies that, independently of the conditions assigned at time $t=0$ (provided $\hat{n}_0 \neq \hat{n}_\pm$), both in the extreme future and in the remote past the boomeron escapes, or recedes, toward the right, with the same asymptotic speed

$$\xi(t) - (\xi_0 + v\,|t|) = 0\left[\exp(-2a\nu_-\,|t|)\right], \quad \text{as } t \to \pm\infty . \qquad (3.47)$$

Note also that the asymptotic speed v (see Eq.(3.45)) is a monotonic function of the dimensionless parameter z (defined by Eq.(3.38)), this parameter being proportional to the width of the soliton; the speed v can take its value only in the finite interval $b\cos\theta \leq v \leq b$, approaching its maximum value b for a very narrow soliton ($\lambda \to 0$) and its minimum value $b\cos\theta$ for a very flat one ($\lambda \to \infty$).

In the special case $\hat{a}.\hat{b} = 0$, a critical value of the width $\lambda$ exists, which is $\lambda_c = 2b/a$ (corresponding to $z=1$), such that a soliton with $\lambda < \lambda_c$ behaves quite differently from a soliton with $\lambda > \lambda_c$. Indeed, if the initial polarization coincides with either one of the unit vectors ($\theta(x) = 1$ if $x > 0$, $\theta(0) = 1/2$, $\theta(x) = 0$ if $x < 0$)

$$\hat{n}_\pm = -z^{-1}\{\pm(z^2-1)^{1/2}\,\theta(z-1)\hat{a} \pm z(1-z^2)^{1/2}\,\theta(1-z)\hat{b} + \left[z^2\,\theta(1-z) + \theta(z-1)\right]\hat{a}\wedge\hat{b}\} \qquad (3.48)$$

it does not change with time, $\hat{n}(t) = \hat{n}_0$, and the soliton does not move at all if $z \geq 1$, while if $z < 1$ it moves with constant speed

$$\xi(t) = \xi_0 \pm vt \ \text{ if } \ \hat{n}_0 = \hat{n}_\pm , \quad v = b(1-z^2)^{1/2} . \qquad (3.49)$$

If instead the initial polarization $\hat{n}_0$ coincides neither with $\hat{n}_+$ nor with $\hat{n}_-$, the polarization does change with time: it never coincides with $\hat{n}_+$ nor $\hat{n}_-$, but, if $z \leq 1$, it tends asymptotically to the values $\hat{n}_\pm$ (i.e. $\hat{n}(\pm\infty) = \hat{n}_\pm$), while if $z > 1$ it precedes periodically, with period

$$T = (2\pi\, z/a)(z^2-1)^{-1/2} . \qquad (3.50)$$

As for the behaviour of the coordinate $\xi(t)$, it is best understood noting that it coincides with that of a particle of unit mass, with initial position $\xi(0) = \xi_0$ and initial speed $\xi_t(0) = -\vec{b}.\hat{n}_0$, moving in the external potential $\tilde{\Phi}(\xi - \xi_0)$ with

$$\tilde{\Phi}(x) = (1/2)b^2 \exp(-2x/\lambda)\,\{\left[([\alpha^2+\gamma^2+z\gamma]^2+z^2\alpha^2)/(\alpha^2+\gamma^2)\right]\exp(-2x/\lambda) - 2z(z+\gamma)\} \tag{3.51}$$

where $\alpha = (\hat{a}.\hat{n}_0)$ and $\gamma = (\hat{a}\wedge\hat{b}.\hat{n}_0)$. The total energy of this particle has the simple expression

$$E = \tilde{\Phi}(0) + (1/2)(\vec{b}.\hat{n}_0)^2 = (1/2)v^2 \,\mathrm{sign}(1-z) \tag{3.52}$$

with v defined by Eq.(3.49). Thus it is positive if $z < 1$, in which case we have a boomeron (see fig.(1)) which escapes, or recedes, to the right as $t \to \pm\infty$, moving asymptotically with the constant speed v. When the total energy is negative, for $z > 1$, this soliton, that we name trappon (see fig.(2)), behaves as a particle trapped in the potential, oscillating indefinitely around the equilibrium position $\xi_0 + x_m$, where $x_m$ is the value of x where the potential (3.51) takes its single negative minimum. This oscillatory motion is periodic with the period given by the formula (3.50). The marginal case $z = 1$ corresponds to the motion of a zero-energy particle that always escapes (or recedes) to positive infinity, but with a vanishing asymptotic velocity. We note, however, that trappons can exist only in this case $(\hat{a}.\hat{b}) = 0$.

It should be emphasized that, for a given SNEE of type (2.45) with $(\hat{a}.\hat{b}) = 0$, in a many-soliton solution all the types of soliton behaviour described above may be simultaneously present, depending on the values of the width $\lambda$ (in particular, whether or not it exceedes the value 2b/a) and of the initial polarization $\hat{n}_0$ of each soliton.

The additional degrees of freedom of our solitons due to their polarization certainly make their dynamics richer than it was for the usual solitons. Indeed, an M-soliton solution will break up, as $t \to \pm\infty$, in $\bar{M} \le M$ bumps with the usual one-soliton shape, $M - \bar{M}$ being therefore the number of trappons contained in this solution. The interaction among these solitons, due to the nonlinearity of the evolution equation, should manifest itself in the asymptotic value of the polarizations of the (asymptotically) free solitons which show up at large $|t|$; in addition to this, one should also find a displacement of the j-th soliton asymptotic position with respect to the undisturbed one-soliton motion, namely, with respect to the function $\xi^{(j)}(t)$ entering into the spectral transform through the spectral parameter $\rho^{(j)}(t)$ according to the formula (3.6). As an example of these processes, we briefly report the main results on the boomeron-boomeron and boomeron-trappon collisions. Let $\{R(k,t) = 0,\ p^{(j)},\ \rho^{(j)}(t),\ P^{(j)}(t),\ j = 1,2\}$ be the ST of the two-soliton solutions.

Assume first that $\xi^{(1)}(t)$, $P^{(1)}(t)$ and $\xi^{(2)}(t)$, $P^{(2)}(t)$ are two boomeron-type solutions of the evolution equations (3.12) and (3.13), and let us introduce their asymptotic behaviour

$$\lim_{t\to\pm\infty}\left[\xi^{(k)}(t)-\xi_{\pm}^{(k)}(t)\right]=0\ ,\ \xi_{\pm}^{(k)}(t)=\xi_{0}^{(k)}\pm v_k t\ ,\ k=1,2 \tag{3.53a}$$

$$\lim_{t\to\pm\infty}P^{(k)}(t)=P_{\pm}^{(k)}\ ,\qquad k=1,2\ . \tag{3.53b}$$

To be definite, let boomeron 1 be asymptotically faster than boomeron 2, i.e. $v_1>v_2>0$. As $t\to\pm\infty$, the two-boomeron solution (see fig.(3)) shows two well separated solitons, which freely move with position $X_{\pm}^{(k)}(t)$ and polarization $\Pi_{\pm}^{(k)}$, each of them being characterized by the usual expression

$$Q_{\pm}^{(k)}(x,t)=-\{A^{(k)}/\cosh^2\{\left[x-X_{\pm}^{(k)}(t)\right]/\lambda^{(k)}\}\}\,\Pi_{\pm}^{(k)} \tag{3.54}$$

where $A^{(k)}=2p^{(k)2}$ and $\lambda^{(k)}=1/p^{(k)}$. The effect of the collision can be described as follows: the asymptotic motion and polarization of the (asymptotically) faster boomeron are not changed at all

$$X_{\pm}^{(1)}(t)=\xi_{\pm}^{(1)}(t)\ ,\quad \Pi_{\pm}^{(1)}=P_{\pm}^{(1)} \tag{3.55}$$

instead, for the (asymptotically) slower boomeron we find an asymptotic displacement

$$X_{\pm}^{(2)}(t)=\xi_{\pm}^{(2)}(t)+d_{\pm},\ d_{\pm}=-(1/2)\lambda^{(2)}\ln\left[1-4\mu^{(1)}\mu^{(2)}\,\mathrm{tr}\,(P_{\pm}^{(1)}\,P_{\pm}^{(2)})\right] \tag{3.56}$$

and the following symptotic polarization

$$\Pi_{\pm}^{(2)}=\exp(2d_{\pm}/\lambda^{(2)})\,\{(\mu^{(1)}/\mu^{(2)})\left[1-\exp(-2d_{\pm}/\lambda^{(2)}\right]P_{\pm}^{(1)}+P_{\pm}^{(2)}-2\mu^{(1)}\{P_{\pm}^{(1)},P_{\pm}^{(2)}\}\} \tag{3.57}$$

where we have set $\mu^{(k)}\equiv p^{(k)}/(p^{(1)}+p^{(2)})$. Specializing this result to the boomeron equation with $(\hat{a}.\hat{b})=0$, we obtain

$$d_{+}=d_{-}\equiv d=-(1/2)\,\lambda^{(2)}\,\ln\{1-2\mu^{(1)}\mu^{(2)}\left[1+z^{(1)}z^{(2)}+(1-z^{(1)2})^{1/2}(1-z^{(2)2})^{1/2}\right]\} \tag{3.58a}$$

$$\Pi_{\pm}^{(2)} = (1/2)(1 + \hat{m}_{\pm}^{(2)}.\vec{\sigma}), \quad \hat{m}_{\pm}^{(2)} = \exp(2d/\lambda^{(2)}) \{ (\mu^{(1)}/\mu^{(2)}) . \left[1 - 2\mu^{(2)} - \exp(-2d/\lambda^{(2)})\right] \hat{n}_{\pm}^{(1)} + (1 - 2\mu^{(1)}) \hat{n}_{\pm}^{(2)} \} \tag{3.58b}$$

with an obvious meaning of the symbols (see Eqs.(3.32), (3.38), (3.48) and (3.49)).

Assume now that the two-soliton solutions of the boomeron equation (with $(\hat{a}.\hat{b}) = 0$) describes a boomeron-trappon system and that $\xi^{(1)}(t)$, $P^{(1)}(t)$ and $\xi^{(2)}(t)$, $P^{(2)}(t)$ are the two solutions of the evolution equations (3.34a,b) corresponding to the boomeron and to the trappon, respectively; namely $z^{(1)} < 1$ and $z^{(2)} > 1$. In this case (see fig.(4)), for large $|t|$, one finds one soliton moving towards or from infinity, and one trappon moving in a confined region. Therefore the asymptotic behaviours (3.53a) and (3.53b) apply now only to the boomeron (i.e. for $k = 1$), while the function $\xi^{(2)}(t)$ and the projection matrix $P^{(2)}(t)$ are periodic functions of t, with period $T = 2\pi\, z^{(2)}/[a(z^{(2)^2} - 1)^{1/2}]$. As $t \to \pm\infty$, we find that the boomeron bump is given by the expression (3.54) (for $k = 1$), with $A^{(1)} = 2p^{(1)^2}$, $\lambda^{(1)} = 1/p^{(1)}$, and its asymptotic motion and polarization are given by the formula (3.55), to say that the collision with the trappon does not affect the asymptotic motion of the boomeron, which therefore behaves for large $|t|$, as if it never met the trappon on its way. On the contrary, the asymptotic motion of the trappon is not that of an undisturbed trappon, namely that characterized by the position $\xi^{(2)}(t)$ and polarization $P^{(2)}(t)$ as for the one-trappon solution discussed above. Now the asymptotic trappon, as $t \to \pm\infty$, is

$$Q_{\pm}^{(2)}(x,t) = -\{ A^{(2)}/\cosh^2 \{ [x - x_{\pm}^{(2)}(t)]/\lambda^{(2)} \} \} \Pi_{\pm}^{(2)}(t) \tag{3.59}$$

with $A^{(2)} = 2p^{(2)^2}$, $\lambda^{(2)} = 1/p^{(2)}$ and

$$x_{\pm}^{(2)}(t) = \xi^{(2)}(t) + \delta_{\pm}(t) \quad ,$$

$$\delta_{\pm}(t) = -(1/2)\lambda^{(2)} \ln\{1 - 4\mu^{(1)} \mu^{(2)} \operatorname{tr}\left[P_{\pm}^{(1)} P^{(2)}(t)\right]\} \tag{3.60a}$$

$$\Pi_{\pm}^{(2)}(t) = \exp\left[2\delta_{\pm}(t)/\lambda^{(2)}\right] \{(\mu^{(1)}/\mu^{(2)})(1 - \exp\left[-2\delta_{\pm}(t)/\lambda^{(2)}\right]) P_{\pm}^{(1)} + P^{(2)}(t) - 2\mu^{(1)} \{ P_{\pm}^{(1)}, P^{(2)}(t)\} \} \, . \tag{3.60b}$$

We notice that the functions $\delta_{\pm}(t)$ and the polarizations $\Pi_{\pm}^{(2)}(t)$ are again periodic in time, with the same period $T = 2\pi\, z^{(2)}/[a(z^{(2)} - 1)^{1/2}]$. Few words are appropriate to comment on these findings. First, we note that the center of mass of the two-soliton system does not freely move as for the KdV equation, because here the solitons, in addition to their interaction, move in their corresponding "external potential" (3.51). The second point concerns the strange asymmetry between the two colliding boomerons, namely boomeron 1 does modify the asymptotic motion of boomeron 2 but not viceversa; the same asymmetry occurs in the boomeron-trappon collision. This feature of the collision processes described by the boomeron equation originates from the asymmetric boundary conditions (3.31) which imply the simple rule: the asymptotic motion and polarization of each soliton is affected only by those other solitons which asymptotically move on its right side, but not by those solitons which asymptotically remain on its left side.

## 4. BASIC NONLINEAR EQUATIONS AND RELATED RESULTS

In sect.2 the basic formulae (2.17) and (2.18) have been investigated in the limit $Q'(x) \to Q(x)$, thus deriving the class of SNEE's. This is just one of several results implied by these basic formulae. Other results, concerning important properties of the solutions of these SNEE's, obtain without taking the limit $Q'(x) \to Q(x)$ and will be tersely presented here. To simplify this presentation and to focus only on the main ideas, throught this section we will avoid the matrix formalism considering only the scalar field case ($N = 1$); the generalization |11| to the matrix case, although essential to investigate important nonlinear evolution equations, does not contribute to our general understanding of this matter.

We focus on the formulae for the reflection coefficients, since they are sufficient to obtain all the interesting final results. The derivation of analogous relationship for the transmission coefficients can be done by a closely analogous technique; we omit also to report on the analogous results for the discrete-spectrum parameters that can be similarly derived by the generalized wronskian method, or from the results given below for the reflection coefficient by assuming the validity of eq.(1.16) (one should be aware, however, that some fine points deserve a more detailed treatment).

Let us first rewrite down the two basic equations (2.17) and (2.18) in this simpler (scalar) case

$$2ikf(-4k^2)\left[R'(k)-R(k)\right] = \int_{-\infty}^{+\infty} dx\, \Psi'(x,k)\, \Psi(x,k)\, \{ f(\Lambda) \left[Q'(x) - Q(x)\right] \} \qquad (4.1)$$

$$(2ik)^2 g(-4k^2) \left[R'(k) + R(k)\right] = \int_{-\infty}^{+\infty} dx\, \Psi'(x,k)\, \Psi(x,k)\, \{ g(\Lambda) \left[Q'_x(x) + Q_x(x) + (Q'(x) - Q(x)) \int_x^{+\infty} dx' (Q'(x')-Q(x'))\right] \} \qquad (4.2)$$

where $Q(x)$ and $Q'(x)$ are real functions vanishing at infinity (see sect.1), and $R(k)$ and $R'(k)$ are their corresponding reflection coefficients; $f(z)$ and $g(z)$ are arbitrary (entire) function and $\Lambda$ is the following linear integro-differential operator depending on $Q(x)$ and $Q'(x)$.

$$\Lambda F(x) = F_{xx}(x) - 2\left[Q'(x)+Q(x)\right]F(x) + \left[Q'_x(x) + Q_x(x)\right]\int_x^{+\infty} dx'\, F(x') + \left[Q'(x) - Q(x)\right]\int_x^{+\infty} dx' \int_{x'}^{+\infty} dx'' \left[Q'(x') - Q(x')\right] F(x'') \quad . \tag{4.3}$$

The basic nonlinear equations obtain directly form eqs.(4.1) and (4.2), since they imply that, if the two functions $Q(x)$ and $Q'(x)$ are related by the formula

$$f(\Lambda)\left[Q'(x) - Q(x)\right] + g(\Lambda)\left[Q'_x(x) + Q_x(x) + (Q'(x) - Q(x))\,.\int_x^{+\infty} dx'\,(Q'(x') - Q(x'))\right] = 0 \tag{4.4}$$

the corresponding reflection coefficients $R(k)$ and $R'(k)$ are related by the equation

$$R'(k) = \{\left[f(-4k^2) - 2ikg(-4k^2)\right]/\left[f(-4k^2) + 2ikg(-4k^2)\right]\}\,R(k) \quad . \tag{4.5}$$

The importance of formula (4.4) is that it displays in closed form the explicit, if complicated, relationship between two potentials $Q(x)$ and $Q'(x)$ whose corresponding reflection coefficients are related by the simple linear formula (4.5).

It should be emphasized that the functions f and g might also depend on other variables (such as t, see below); they must of course be independent of x.

We now use these important findings as a tool to investigate the properties of the solution of a SNEE of the class (3.1) with the initial condition

$$Q(x,0) = Q_0(x) \tag{4.6}$$

eq.(3.1) obtains, of course, from (2.28) for $N = 1$, or, it can be directly derived from our basic equation (4.4) by choosing $Q'(x) = Q(x,t+\Delta t)$, $Q(x) = Q(x,t)$, $f(z) = 1/\Delta t$, $g(z) = -\beta_0(z)$ and taking the limit $\Delta t \to 0$. The linear operator L is defined here as

$$LF(x) = F_{xx}(x) - 4Q(x,t)\,F(x) + 2Q_x(x,t)\int_x^{\infty} dx' F(x') \quad . \tag{4.7}$$

Let us regard then solution of the Cauchy problem (3.1) and (4.6) as a "potential" which depends parametrically on the time variable t, on the initial condition $Q_0(x)$ and on the function $\beta_0(z)$ characterizing a particular SNEE

$$Q(x) = Q(x,t,Q_0,\beta_0) \quad . \tag{4.8}$$

A systematic application of the basic equations (4.4) and (4.5) is based on considering Q' as obtained from Q by changing only t (the result being a resolvent formula), or by changing only $Q_0(x)$ (the result being a Bäcklund transformation) or by changing only $\beta_0(z)$. More general transformations are obtained |27| combining together these three transformations.

i) Bäcklund transformations |28|: Assume that in eq.(4.4)

$$Q'(x) = Q(x,t,Q_0',\beta_0) \; , \quad Q(x) = Q(x,t,Q_0,\beta_0) \tag{4.9}$$

then (applying our golden rule) let us write the relevant equations in the k-space, namely

$$R(k,t) = \exp\left[4ik\beta_0\,(-4k^2)t\right] R_0(k) \tag{4.10}$$

$$R'(k,t) = \left\{ \left[f(-4k^2) - 2ikg(-4k^2)\right] / \left[f(-4k^2) + 2ikg(-4k^2)\right] \right\} R(k,t) \; ; \tag{4.11}$$

therefore from these equations it is clear that the assumption (4.9) is verified if the functions f and g are independent of t. Indeed, in this case, R'(k,t) is related to $R_0'(k) \equiv R'(k,0)$ by the same equation (4.10) that relates R(k,t) to $R_0(k) \equiv R(k,0)$. But such a time evolution for the reflection coefficient R'(k,t) corresponds to the time evolution (3.1) for the corresponding potential Q'(x,t). Conclusion: if Q(x,t) satisfies the SNEE (3.1), and Q'(x,t) is related to Q(x,t) by (4.4), with f and g independent of t, then Q'(x,t) satisfies the same SNEE (3.1).

This class of Bäcklund transformations is interesting because it holds for the whole class (3.1) of SNEE's, there being no restriction on the functions f and g that characterize the Bäcklund transformation. The condition that a real solution Q(x,t) should be transformed in a real solution Q'(x,t) by (4.4) is just the requirement that f(z) and g(z) be real analytic:

$$f(z) = f^*(z^*) \; , \; g(z) = g^*(z^*) \; . \tag{4.12}$$

A restriction on f and g is, however, implied by the requirement that both Q and Q' have the properties that were assumed to begin with; in particular the property that, if the function Q(x) is finite valued for real x and vanishes asymptotically (as $x \to \pm\infty$), the function Q'(x), related to Q(x) by (4.4), also has these properties; but, while it would be difficult to specify such a restriction on the basis of eq.(4.4), it is actually quite easy to interpret it on the basis of eq. (4.5), since it then amounts to the requirement that the simple multiplicative factor

$$\phi(k) = \left[f(-4k^2) - 2ikg(-4k^2)\right] / \left[f(-4k^2) + 2ikg(-4k^2)\right] \tag{4.13}$$

does not bring about any unacceptable property of the reflection coefficient R'(k). This means, for instance, that $\phi(k)$ (that is generally a meromorphic function of k) should not have poles of higher order than the first, or even simple poles coinciding with poles of R(k), or poles in disallowed regions fo the complex k-plane (such as the region Re $k \neq 0$, Im $k > 0$, if the potentials are required to be real and to vanish asymptotically faster than exponentially).

Let us proceed here to an analysis of the simpler Bäcklund transformation, that corresponds to both f and g being constants. It is convenient to characterize this transformation by the constant $p = \frac{1}{2} f/g$, so that (4.4) reduces to the one-parameter subclass of transformations

$$R'(k,t) = -\left[(k+ip)/(k-ip)\right] R(k,t) \tag{4.14}$$

and the Bäcklund transformation reads

$$Q'(x,t) = Q(x,t) - (2p)^{-1}\left[Q'_x(x,t) + Q_x(x,t) + (Q'(x,t) - Q(x,t)) \cdot \int_x^{+\infty} dx'(Q'(x',t) - Q(x',t))\right] . \tag{4.15}$$

To discuss these Bäcklund transformations and related results, it is convenient to work in terms of the integral of Q, rather than Q itself. We therefore introduce the function

$$W(x,t) = \int_x^{\infty} dx' Q(x',t) , \quad W'(x,t) = \int_x^{\infty} dx' Q'(x',t) \tag{4.16}$$

that clearly satisfy the boundary conditions

$$W(+\infty,t) = W_x(\pm\infty,t) = W'(+\infty,t) = W'_x(\pm\infty,t) = 0 \tag{4.17}$$

and from which the potentials can of course be recovered through the formula

$$Q(x,t) = - W_x(x,t) \ , \quad Q'(x,t) = - W'_x(x,t) \ . \tag{4.18}$$

Indeed it is often convenient to write also the SNEE's (2.28) in terms of W rather than Q; for instance, it is in terms of W (compare (2.46) with (4.16)) that the boomeron equation |20| (2.45) can be written as a pure partial differential, rather than integro-differential, nonlinear equation.

In terms of W and W' the Bäcklund transformation (4.15) becomes

$$W'_x(x,t) + W_x(x,t) = - \frac{1}{2} \left[W'(x,t) - W(x,t)\right] \left[4p + W'(x,t) - W(x,t)\right] \ . \tag{4.19}$$

Clearly this equation is invariant under the transformation $W \longleftrightarrow W'$, $p \longleftrightarrow -p$; and this is consistent with the corresponding transformation (4.14) in k-space, that is clearly invariant under $R \longleftrightarrow R'$, $p \longleftrightarrow -p$. Thus the Bäcklund transformation with parameter $-p$ can be considered the inverse transformation to that with parameter p. Note, however, that generally (4.19) with a given W (consistent with (4.17)) can be integrated to yield a W' consistent with (4.17), only if $p > 0$; and indeed in such a case the boundary conditions (4.17) are satisfied automatically, so that there still remains a certain arbitrariness in W', since the "constant of integration" is not fixed by the boundary conditions (4.17). Note that this "constant of integration" does not appear in the transformation (4.14) of the reflection coefficient and therefore it must be related to the discrete (soliton) component of the ST of $Q'(x,t)$; consistently this "constant of integration" implies a dependence on t, that is different for the different equations of the class (3.1) and that can be ascertained only by inserting Q' in the SNEE itself, thereby obtaining an equation, involving only the time variable, for the "integration constant". This situation is clearly connected to the fact that the formula (4.19) is often referred in the literature as "one half" of a Bäcklund transformation; for the derivation within our formalism of the other "one half", for the KdV equation, see K.M. Case and S.C. Chiu |29|. Therefore, the picture implied by (4.14), and our previous analysis, is that, by solving (4.19) for W', one gets generally a solution $Q' = - W'_x$ having one more soliton (corresponding to the discrete eigenvalue $k^2 = -p^2$), than the solution $Q = -W_x$ (soliton creation); this also implies that, for a given Q having a discrete eigenvalue for $k = iq$, $q > 0$, eqs.(4.19) and (4.17) would exceptionally be solvable for W' even for a negative value of p, namely for $p = -q$; the corresponding solution Q' would then have one less soliton that Q (soliton annihilation).

It is instructive to solve eq.(4.19) in the special case $W = 0$. One then obtains for $W(x,t)$, the single soliton solution (3.2), namely

$$W'(x,t) = -2p\,\{1 - \mathrm{tgh}\,[p(x-\xi)]\} \tag{4.20}$$

the constant of integration $\xi$ depends of course on t; its explicit time evolution may be ascertained by inserting (4.20) in the SNEE (3.1) and by solving the resulting equation, that involves only the variable t, recovering thereby for $\xi(t)$ the expression (3.4).

An important implication of the expression (4.5) is that all Bäcklund transformations (4.4) (and therefore the subclass of them (4.15), or, equivalently, (4.19)) commute; it should be emphasized that this property, that is essentially trivial when interpreted through (4.5), is instead highly nontrivial when viewed in the context of (4.4) (or even (4.19)), due to the nonlinear structure of these formulae. An important consequence of this property is the so-called "nonlinear superposition principle". This can be derived in the following way: let $Q_o(x,t)$ be a solution of SNEE (3.1), $Q_1(x,t)$ respectively $Q_2(x,t)$, the solution of the same SNEE related to it by the Bäcklund transformation (4.15) with $p = p^{(1)}$, respectively $p = p^{(2)}$, $Q_{12}(x,t)$ the solution related to $Q_1(x,t)$ by the Bäcklund transformation (4.15) with $p = p^{(2)}$, and $Q_{21}(x,t)$ the solution related to $Q_2(x,t)$ by the Bäcklund transformation (4.15) with $p = p^{(1)}$. Then, with obvious notation, we get from (4.14)

$$R_{12}(k,t) = \{[(k+ip^{(2)})(k+ip^{(1)})]/[(k-ip^{(2)})(k-ip^{(1)})]\}\,R_o(k,t) \tag{4.21a}$$

$$R_{21}(k,t) = \{[(k+ip^{(1)})(k+ip^{(2)})]/[(k-ip^{(1)})(k-ip^{(2)})]\}\,R_o(k,t) \tag{4.21b}$$

that clearly imply

$$R_{12}(k,t) = R_{21}(k,t) \tag{4.22}$$

and therefore also

$$Q_{12}(x,t) = Q_{21}(x,t)\quad . \tag{4.23}$$

Let us write out the formulae corresponding to the statements that we have just made, working again with the more convenient quantities W

$$W_{1x} + W_{ox} = -(1/2)(W_1 - W_o)(4p^{(1)} + W_1 - W_o) \tag{4.24a}$$

$$W_{2x} + W_{ox} = -(1/2)(W_2 - W_o)(4p^{(2)} + W_2 - W_o) \tag{4.24b}$$

$$W_{12x} + W_{1x} = -(1/2)(W_{12} - W_1)(4p^{(2)} + W_{12} - W_1) \tag{4.24c}$$

$$W_{12x} + W_{2x} = -(1/2)(W_{12} - W_2)(4p^{(1)} + W_{12} - W_2) \tag{4.24d}$$

where we have taken into account that $W_{12} = W_{21}$, as implied by (4.23) and (4.16). With a little algebra, we now eliminate all differential terms, getting thereby the formula

$$W_{12}(x,t) = W_0(x,t) - 2(p^{(1)} + p^{(2)}) \left[W_1(x,t) - W_2(x,t)\right] \cdot \left[2(p^{(1)} - p^{(2)}) + W_1(x,t) - W_2(x,t)\right]^{-1} \qquad (4.25)$$

that provides, through (4.16), an explicit expression of the solution $Q_{12}$ of the SNEE (3.1), in terms of an arbitrary solution $Q_0$ and of the two solutions $Q_1$ and $Q_2$ related to $Q_0$ by the simple Bäcklund transformation (4.15).

Among the implications of a formula such as (4.25), we merely mention here that generally $Q_{12}$ has two more solitons than $Q_0$, and that, starting from $Q_0 = 0$, as shown first, for the KdV equation, by Wahlquist and Estabrook |31|, one can use (4.25) to generate the whole ladder of multisoliton solutions. For instance, it can be verified that the two-soliton solutions

$$W(x,t) = -2(p^{(1)} + p^{(2)})(1 - \tau_1\tau_2)^{-1}(\tau_1 + \tau_2 - 2\tau_1\tau_2) \qquad (4.26a)$$

$$\tau_k \equiv p^{(k)}(p^{(1)} + p^{(2)})^{-1}\{1 - \text{tgh}\{p^{(k)}\left[x - \xi^{(k)}(t)\right]\}\}, \quad k = 1,2, \qquad (4.26b)$$

which can be easily obtained solving the Marchenko equation (1.35) with $R(k,t) = 0$ and $\rho^{(k)}(t) = 2p^{(k)}\exp[2p^{(k)}\,\xi^{(k)}(t)]$, $k = 1,2$, is recovered by inserting in (4.25), together with $W_0 = 0$, the two solutions

$$W_1(x,t) = -2p^{(1)}\{1 - \text{tgh}\{p^{(1)}\left[x - \xi^{(1)}(t)\right] + \delta\}\} \qquad (4.27)$$

$$W_2(x,t) = -2p^{(2)}\{1 - \text{cotgh}\{p^{(2)}\left[x - \xi^{(2)}(t)\right] + \delta\}\} \qquad (4.28)$$

where

$$\delta = (1/2)\ \ln\left[(p^{(2)} + p^{(1)})/(p^{(2)} - p^{(1)})\right] \qquad (4.29)$$

and

$$p^{(2)} > p^{(1)} > 0 \ . \qquad (4.30)$$

Note that it is this last inequality that distinguishes the different roles played by the solutions $W_1$ and $W_2$; indeed it can be easily shown that only the choice (4.27) and (4.28) yields a nonsingular solution $W_{12}$.

Since all the results we have given can be extended in a straightforward way to the matrix fields |11|, we will now briefly discuss, as a second remarkable appli_ cation of the Bäcklund transformations, the conservation laws which are satisfied by the solutions of the boomeron equation |19| (2.45). The technique described here to obtain the conserved quantities is closely analogous to the results first given for the KdV equation |31|.

We first note that the scalar equation (2.45a) has already the form of a conservation law. Then we show how to obtain from it an infinite sequence of other conservation laws by exploiting the dependence on the parameter p of the new field variables U' and $\vec{V}'$, obtained from U and $\vec{V}$ by a Backlund transformation, of the type discussed above, characterized by p.

As implied by the formula (4.14), which holds also in the matrix case, the unit transformation is obtained in the limit $p \to \infty$; therefore, in order to discuss a Bäcklund transformation in the neighbour of the identity transformation, it is more convenient to introduce instead the parameter

$$\varepsilon = -(2p)^{-1} . \tag{4.31}$$

The expression for this one-parameter family of Bäcklund transformations is found to be

$$(U'-U)\left[1-(\varepsilon/2)(U'-U)\right] = \varepsilon\left[U'_x + U_x + (1/2)(\vec{V}'-\vec{V})^2\right] \tag{4.32a}$$

$$(\vec{V}'-\vec{V})\left[1-\varepsilon(U'-U)\right] = \varepsilon(\vec{V}'_x + \vec{V}_x) \tag{4.32b}$$

But, of course, also U' and $\vec{V}'$ satisfy the conservation law (2.45a)

$$U'_t(x,t) = \vec{b}.\vec{V}'_x(x,t) \tag{4.33}$$

which, on the other hand, contains a parametric dependence on $\varepsilon$, as implied by (4.32). Introducing now in the Backlund transformation (4.32) the asymptotic expansion

$$U' = U + \sum_{n=1}^{N} \varepsilon^n U^{(n)} + O(\varepsilon^{N+1}) \tag{4.34a}$$

$$\vec{V}' = \vec{V} + \sum_{n=1}^{N} \varepsilon^n \vec{V}^{(n)} + O(\varepsilon^{N+1}) \tag{4.34b}$$

one obtains the recursion relations

$$U^{(n+1)} = U_x^{(n)} + (1/2)\sum_{m=1}^{n-1} (U^{(m)}U^{(n-m)} + \vec{V}^{(m)}\vec{V}^{(n-m)}) \quad , \quad n \geq 1 \tag{4.35a}$$

$$\vec{V}^{(n+1)} = \vec{V}_x^{(n)} + \sum_{m=1}^{n-1} U^{(m)} \vec{V}^{(n-m)} \quad , \quad n \geq 1 \tag{4.35b}$$

with the initial conditions

$$U^{(1)} = 2U_x \, , \quad \vec{V}^{(1)} = 2\vec{V}_x \, . \tag{4.36}$$

It is easily seen that these relations, together with (3.31), imply

$$U^{(n)}(\pm\infty,t) = 0 \, , \quad \vec{V}^{(n)}(\pm\infty,t) = 0 \, , \quad n \geq 1 \, . \tag{4.37}$$

Inserting now (4.34) in (4.33), we obtain the infinite sequence of conservation laws

$$U_t^{(n)}(x,t) = \vec{b}.\vec{V}_x^{(n)}(x,t) \, , \quad n \geq 1 \tag{4.38}$$

that, together with (4.37), imply that the quantities

$$C_n = (1/4) \int_{-\infty}^{+\infty} dx \, U^{(n)}(x,t), \; n \geq 1 \tag{4.39}$$

are constants of the motion. It turns, however, out that the quantities $U^{(2n)}(x,t)$ are perfect derivatives of quantities that vanish asymptotically, so that $C_{2n} = 0$. The odd-numbered $C_{2n+1}$ do instead yield nontrivial conserved quantities, for instance:

$$C_1 = -(1/2) \, U(-\infty,t) \tag{4.40a}$$

$$C_3 = (1/2) \int_{-\infty}^{+\infty} dx \left[U_x^2(x,t) + \vec{V}_x^2(x,t)\right] \tag{4.40b}$$

$$C_5 = \int_{-\infty}^{+\infty} dx \{U_x(x,t)\left[U_x^2(x,t) + 3\vec{V}_x^2(x,t)\right] - (1/2)\left[U_{xx}^2(x,t) + \vec{V}_{xx}^2(x,t)\right]\} \, . \tag{4.40c}$$

It can be easily verified that these constants of the motion have a simple expression in terms of the 2 x 2 matrix field $Q(x,t)$ related to the scalar and vector fields $U(x,t)$ and $\vec{V}(x,t)$ by (2.46); for instance, the expressions (4.40) then read

$$C_1 = -\,(1/4) \; \mathrm{tr} \left[\int_{-\infty}^{+\infty} dx \, Q(x,t)\right] \tag{4.41a}$$

$$C_3 = (1/4) \mathrm{tr} \, \{ \int_{-\infty}^{+\infty} dx \left[Q(x,t)\right]^2 \} \tag{4.41b}$$

$$C_5 = -(1/4) \; \mathrm{tr} \, \{ \int_{-\infty}^{+\infty} dx \, \{ 2 \left[Q(x,t)\right]^3 + \left[Q_x(x,t)\right]^2 \} \} \, . \tag{4.41c}$$

For the single-soliton solution (3.33), or (3.11), these conserved quantities take the simple expressions $C_1 = \lambda^{-1}$, $C_3 = (4/3)\,\lambda^{-3}$, $C_5 = (16/5)\,\lambda^{-5}$.

ii) Resolvent formula: Assume |18| now that in eq.(4.4)

$$Q'(x) = Q(x,t,Q_0,\beta_0) \quad , \quad Q(x) = Q(x,0,Q_0,\beta_0) \tag{4.42}$$

then in the k-space eq.(4.5) should be read with

$$R'(k) = R(k,t) \; , \quad R(k) = R(k,0) = R_0(k) \; . \tag{4.43}$$

But then eq. (4.5) becomes consistent with the actual time evolution (4.10) if we choose

$$f(z^2,t) = \cosh\left[z\,\beta_0(z^2)t\right] \; , \quad g(z^2,t) = -\,z^{-1}\sinh\left[z\,\beta_0(z^2)t\right] \; . \tag{4.44}$$

Therefore the corresponding transformation in x-space

$$\cosh\left[\Lambda^{1/2}\,\beta_0(\Lambda)t\right]\left[Q(x,t)-Q_0(x)\right]-\Lambda^{-1/2}\sinh\left[\Lambda^{1/2}\beta_0(\Lambda)t\right]\{\,Q_x(x,t)+ Q_{0x}(x)+\left[Q(x,t)-Q_0(x)\right]\int_x^{+\infty}dx'\left[Q(x',t)-Q_0(x')\right]\} = 0 \tag{4.45}$$

relates a function $Q_0(x)$ just to the function $Q(x,t)$ into which $Q_0(x)$ has evolved at time t according to the SNEE (3.1) (the initial condition (4.6) being trivially verified by setting $t=0$ in (4.45), and $\Lambda$ being the operator (4.3) with $Q'(x)$ and $Q(x)$ replaced by $Q(x,t)$ and $Q_0(x)$ respectively). For this reason we name the functional equation (4.45) the resolvent formula.

An interesting, if complicate, operator identity obtains by choosing

$$2\beta_0(z) = v \quad , \quad vt = a \quad , \quad Q_0(x) = f(x) \tag{4.46}$$

therefore, solving eq.(3.1),

$$Q(x,t) = f(x+a) \tag{4.47}$$

and (4.45) becomes the following operator identity

$$\cosh\,(\Lambda^{1/2}\,a/2)\left[f(x+a)-f(x)\right] - \Lambda^{-1/2}\sinh\,(\Lambda^{1/2}\,a/2)\{\,f_x(x+a)+f_x(x)+ \left[f(x+a)-f(x)\right]\int_x^{+\infty}dx'\left[f(x'+a)-f(x')\right]\} = 0 \tag{4.48}$$

where f(x) is an arbitrary function, except for the requirement that it vanishes asymptotically, and $\Lambda$ is the operator

$$\Lambda F(x) = F_{xx}(x) - 2\left[f(x+a) + f(x)\right] F(x) + \left[f_x(x+a) + f_x(x)\right] \int_x^{+\infty} dx' \, F(x') \; +$$

$$\left[f(x+a) - f(x)\right] \int_x^{+\infty} dx' \int_{x'}^{+\infty} dx'' \left[f(x'+a) - f(x')\right] F(x'') \; . \tag{4.49}$$

Note that the linearization of this equation yields the elementary operator identity

$$f(x+a) = \exp(a\, d/dx)\, f(x) \; . \tag{4.50}$$

iii) Transformation between solutions of different SNEE's (of the same class): Assume finally that in eq.(4.4)

$$Q'(x) = Q(x,t,Q_0,\beta_0') \equiv Q'(x,t) \quad , \quad Q(x) = Q(x,t,Q_0,\beta_0) \equiv Q(x,t) \tag{4.51}$$

which, of course, implies $Q'(x,0) = Q(x,0) = Q_0(x)$; then, the corresponding reflection coefficients $R'(k,t)$ and $R(k,t)$, that evolve in time as

$$R'(k,t) = \exp\left[4ik\, \beta_0'(-4k^2)t\right] R_0(k) \tag{4.52a}$$

$$R(k,t) = \exp\left[4ik\, \beta_0(-4k^2)t\right] R_0(k) \tag{4.52b}$$

where $R_0(k) = R(k,0) = R'(k,0)$, are related to each other by the transformation (4.5), $R'(k)$ and $R(k)$ being replaced in this equation by $R'(k,t)$ and $R(k,t)$ given by (4.52a) and (4.52b) respectively. The transformation (4.5) is then consistent with the relationship which obtains by eliminating $R_0(k)$ from (4.52), namely

$$R'(k,t) = \exp\left\{4ik\left[\beta_0'(-4k^2) - \beta_0(-4k^2)\right]t\right\} R(k,t) \tag{4.53}$$

if we whoose

$$f(z^2,t) = \cosh\{z\left[\beta_0'(z^2) - \beta_0(z^2)\right]t\} \; , \; g(z^2,t) = -z^{-1}\sinh\{z\left[\beta_0'(z^2) - \beta_0(z^2)\right]\} . \tag{4.54}$$

Therefore eqs.(4.53), (4.52) and (4.54) imply that, if $Q(x,t)$ satisfies the SNEE (3.1) with the initial condition (4.6), then $Q'(x,t)$ which is related to $Q(x,t)$ by the transformation

$$\cosh\{\Lambda^{1/2}\left[\beta_0'(\Lambda)-\beta_0(\Lambda)\right]t\}\left[Q'(x,t)-Q(x,t)\right]-\Lambda^{-1/2}\sinh\{\Lambda^{1/2}\left[\beta_0'(\Lambda)-\beta_0(\Lambda)\right]t\}$$

$$\{Q_x'(x,t)+Q_x(x,t)+\left[Q'(x,t)-Q(x,t)\right]\int_x^{+\infty}dx'\left[Q'(x',t)-Q(x',t)\right]\}=0 \tag{4.55}$$

satisfies the SNEE

$$Q_t'(k) = 2\beta_0'(L')\, Q_x'(x,t) \tag{4.56}$$

with the same initial condition (4.6), $Q'(x,0) = Q_0(x)$; of course, the operator L' is defined by (4.7) with $Q(x,t)$ replaced by $Q'(x,t)$.

For sake of completeness, we now show how to deal also with the translations of the space variable x, within the present scheme based on the transformations (4.4). We first, notice that, if $R(k)$ is the reflection coefficient corresponding to the potential $Q(x)$, then

$$R'(k) = \exp(2ik\,\Delta x)\,R(k) \tag{4.57}$$

is the reflection coefficient corresponding to the potential

$$Q'(x) = Q(x+\Delta x)\quad . \tag{4.58}$$

This result can be easily derived from the Schrödinger equation (1.9) and from the definition (1.10a) of the reflection coefficient. Comparing now (4.57) and (4.5) shows that, to make (4.57) coincide with the transformation (4.5), we can choose

$$f(z^2) = \cosh(z\,\Delta x/2)\ ,\qquad g(z^2) = -z^1 \sinh(z\,\Delta x/2) \tag{4.59}$$

and therefore the corresponding transformation (4.4) implies that $Q(x+\Delta x)$ is related to $Q(x)$ by the operator identity (4.48) we have already derived (with $f(x)$ replaced by $Q(x)$ and a replaced by $\Delta x$).

It should be realized that more general transformations (some of them being reported by F. Calogero and A. Degasperis |27,11|),can be obtained by combining two, or more, of the four types of transformations we have discussed, namely the Bäcklund transformation, the resolvent formula, the transformation (4.55) and the x-translation (4.48). As an example, we now write down the generalized resolvent formula which relates $Q(x+\Delta x,\ t+\Delta t)$ to $Q(x,t)$, $Q(x,t)$ being a solution of the SNEE (3.1),

$$\cosh \{ \Lambda^{1/2} \left[ (\Delta x/2) + \Delta t \; \beta_0(\Lambda) \right] \} \left[ Q(x+\Delta x,\; t+\Delta t) - Q(x,t) \right] -$$

$$\Lambda^{-1/2} \sinh \{ \Lambda^{1/2} \left[ (\Delta x/2) + \Delta t \; \beta_0(\Lambda) \right] \} \{ Q_x(x+\Delta x,\; t+\Delta t) + Q_x(x,t) \; +$$

$$\left[ Q(x+\Delta x,\; t+\Delta t) - Q(x,t) \right] \int_x^{+\infty} dx' \left[ Q(x'+\Delta x,\; t+\Delta t) - Q(x',t) \right] \} = 0 \qquad (4.60)$$

where, of course, $\Lambda$ is the integro-differential operator defined by (4.3) with $Q'(x) = Q(x+\Delta x,\ t+\Delta t)$ and $Q(x) = Q(x,t)$. The interested reader will find the derivation of this equation in the reference |27|.

REFERENCES

1. Gardner, C.S., Greene, J.M., Kruskal, M.D., Miura, R.M.: Method for Solving the Korteweg-de Vries equation. Phys. Rev. Lett. 19, 1095 (1967).

2. Ablowitz, M.J., Kaup, D.J., Newell, A.C., Segur, H.: The Inverse Scattering transform-Fourier analysis for nonlinear problems. Appl. Math. 53, 249 (1974); hereafter referred to as AKNS.

3. Scott, A.C., Chu, F.Y.F., McLaughlin, D.W.: The soliton: a new concept in applied science. Proc. I.E.E.E. 61, 1443 (1973).
Bollough, R.K.: Solitons. In Interaction of Radiation with Condensed Matter. Vol.I, IAEA-SMR- 20/51, Vienna (1977); also lectures delivered at the International Advanced study Institute on Nonlinear Equations in Physics and Mathematics, Istanbul, August 1-13, 1977. Proceedings edited by A.O.Barut.

4. Calogero, F., Degasperis, A.: Solution by the spectral transform method of a nonlinear evolution equation including as a special case the cylindrical KdV equation. Lett. Nuovo Cimento 23, 150 (1978).

5. Well known linear problems are the Schrödinger and the generalized (non selfadjoint) Zackarov-Shabat spectral problems on the real line, with a potential which vanishes at infinity. The most interesting associated evolution equations are the KdV for the first one and the MKdV (modified KdV), sine-Gordon and nonlinear Schrödinger equations for the second one. A unified treatment of these evolution equations can be obtained by considering the N x N matrix Schrödinger spectral problem: see Ref. |12| and Jaulent, M., Miodek, I.: Connection between Zackarov Shabat and Schrödinger Type Inverse Scattering Transforms. Preprint PM/77/9, University of Montpellier (1977). Calogero, F., Degasperis, A.: in preparation. The Schrödinger problem with a potential which depends in a simple way on the eigenvalue has been discussed by Jaulent, M. and Miodek, I.: Nonlinear evolution equations associated with "energy-dependent Schrödinger potentials". Lett. Math. Phys. 1, 243 (1976). The Schrödinger problem with a potential which diverges at infinity has been discussed by Kulish, P.: Inverse scattering problem for Schrödinger equation on a line with potential growing in one direction. Mathematical Notes, Leningrad (1970). Also Calogero, F., Degasperis, A.: Inverse spectral problem for the one-dimensional Schrödinger equation with an additional linear potential. Lett. Nuovo Cimento, 23, 143 (1978).
See also Ref. |4|.

6. Lax, P.D.: Integrals of Nonlinear equations of evolution and solitary waves. Comm. Pure Appl. Math. 21, 467 (1968).

7. Calogero, F.: A Method to Generate Solvable Nonlinear Evolution Equations. Lettere Nuovo Cimento, 14, 443 (1975).
Calogero, F., Degasperis, A.: Nonlinear Evolution Equations Solvable by the Inverse Spectral Transform. I. Nuovo Cimento 32B, 201 (1976).

8. Wahlquist, H.D., Estabrook, F.B.: Prolungation structures of nonlinear evolution equations. J. Math. Phys. 16, 1 (1975).

9. Corones, J., Markovski, B.L., Rizov, V.A.: Bilocal Lie Groups and Solitons. Phys. Lett. 61A, 439 (1977).

10. Novikov, S.P.: New applications of algebraic geometry to nonlinear equations and inverse problems. To appear in the Proceedings of a Symposium held at the Accademia Nazionale dei Lincei in Rome in June 1977, Calogero, F. (editor): Nonlinear evolution equations solvable by the spectral transform. Pitman. London (1978).

11. Calogero, F., Degasperis, A.: Nonlinear evolution equations solvable by the inverse spectral transform.II. Nuovo Cimento 39B, 1 (1977).

12. Wadati, M., Kamijo, T.: On the extension of inverse scattering method. Prog. Theor. Phys. 52, 397 (1974).

13. Newton, R.G.: Scattering Theory of Waves and Particles. Mc Graw Hill Book Company, New York (1966).

14. It can be easily proved that if Q(x) is hermitian and k is real, then T(k) is not singular.

15. Gel'fand, I.M., Levitan, B.M.: On the determination of a differential equation from its spectral function. Amer. Math. Soc. Transl., 1, 253 (1955).
Agranovich, Z.S., Marchenko, V.A.: The Inverse Problem of Scattering Theory. (translated from Russian by B.D. Seckler). New York, Gordon and Breach (1963).
Chadan, K., Sabatier, P.C.: Inverse Problems in Quantum Scattering Theory. Springer Verlag, New York (1977).

16. For a detailed treatment (in the case $N=1$) see the paper by Calogero, F.: Generalized wronskian relations, one-dimensional Schrödinger equation and nonlinear partial differential equations solvable by the inverse-scattering method. Nuovo Cimento 31B, 229 (1976).

17. Indeed this result applies to all SNEE's associated to the generalized Zacharov-Shabat spectral problem since they are a subclass of the present class of SNEE (see below and the references reported in footnote |5|). Nonlinear evolution equations which are not isospectral flows but can be still investigated by the ST method associated to the single-channel Schrödinger problem and the generalized Zackarov-Shabat problem, have been discussed by Newell, A.: The general structure of integrable evolution equations, to appear in Proc. Roy. Soc., 1978; and by Calogero, F., Degasperis, A.: Extension of the spectral transform method for solving nonlinear evolution equations. I & II. Lett Nuovo Cimento 22, 131 and 263 (1978).

18. Since the evolution equation (2.28) is invariant under time-translations, there is no loss of generality in considering $t=0$ as the initial time.

19. Calogero, F., Degasperis, A.: Coupled nonlinear evolution equations solvable via the inverse spectral transform and solitons that come back: the boomeron. Lettere Nuovo Cimento 16, 425 (1976).
Calogero, F., Degasperis, A.: Bäcklund transformations, nonlinear superposition principle, multisoliton solutions and conserved quantities for the "boomeron" nonlinear evolution equation. Lettere Nuovo Cimento 16, 434 (1976).

20. A special class of non hermitian potentials has been discussed by: Bruschi,M. Levi, D., Ragnisco, O.: Evolution equations associated to the triangular matrix Schrödinger problem solvable by the inverse spectral transform. Nuovo Cimento 45A, 225 (1978).

21. Scott-Russel, J.: Report on waves. Report of the Fourteenth Meeting on the British Association for the Advancement of Science, London, 1845, pp.311-390.

22. Zabusky, N.J. Kruskal, M.D.: Interaction of solitons in a collisionless plasma and the recurrence of initial states. Phys. Rev. Lett. 15, 240 (1965)
Zabusky, N.J.: A synergetic approach to problems of nonlinear dispersive wave propagation and interaction. In Nonlinear Partial Differential Equations Ames, W. (editor). New York: Academic Press (1967), pp.223-258.

23. Kruskal, M.D.: The birth of the soliton. Proceedings of the Symposium held at the Accademia Nazionale dei Lincei in Rome in June 1977. Calogero, F. (editor): Nonlinear evolution equations solvable by the spectral transform. Pitman. London (1978).

24. Zackarov, V.E.: Kinetic equation for solitons. Soviet Phys. JETP, 33, 538 (1971).
Wadati, M., Toda, M.: The exact N-soliton solution of the Korteweg-de Vries equation. J.Phys. Soc. Japan 32, 1403 (1972).
Tanaka, S.: Publ. Res. Inst. Math. Sci. Kyoto University 8, 419 (1972/73).
See also: Gardner, C.S., Greene, J.M., Kruskal, M.D., Miura, R.M.: Korteweg-de Vries equation and generalization. VI. Methods for exact solution. Comm. Pure Appl. Math. 27, 97 (1974).

25. Condrey, P.J., Eilbeck, J.C., Gibbon, J.D.: The Sine-Gordon equation as a model classical field theory. Nuovo Cimento 25B, 497 (1975).

26. In the special case $\vec{a} \wedge \vec{b} = 0$ the nonlinear evolution equation (3.34a) coincides with the Landau-Lifshitz equation describing the magnetization of a ferromagnetic substance.

27. Calogero, F., Degasperis, A.: Transformations between solutions of different nonlinear evolution equations solvable via the same inverse spectral transform, generalized resolvent formulas and nonlinear operator identities. Lettere Nuovo Cimento 16, 181 (1976).

28. See, for instance: Lamb Jr., G.L.: Bäcklund transformations for certain nonlinear evolution equations.J. Math. Phys. 15, 2157 (1974);
Chen, H.H.: General derivation of Bäcklund transformations from inverse scattering problems. Phys. Rev. Lett. 33, 925 (1974).
Miura, R.M. (editor) Bäcklund transformations. Lectures Notes in Mathematics, 515, Berlin, Heidelberg, New York: Springer Verlag (1976). See also Ref. |30|.

29. Case, K.M., Chu, S.C.: Some remarks on the wronskian technique and the inverse scattering transform. J. Math. Phys. 18, 2044 (1977).

30. Wahlquist, H.D., Estabrook, F.B.: Bäcklund transformations for solutions of the Korteweg-de Vries Equation. Phys. Rev. Lett. 31, 1386 (1973).

31. See, for instance: Kruskal, M.D.: Nonlinear wave equations, in Dynamical Systems, Theory and Applications. Moser, J. (editor). Lectures Notes in Physics, 38, Berlin, Heidelbcrg, New York: Springer Verlag (1975).

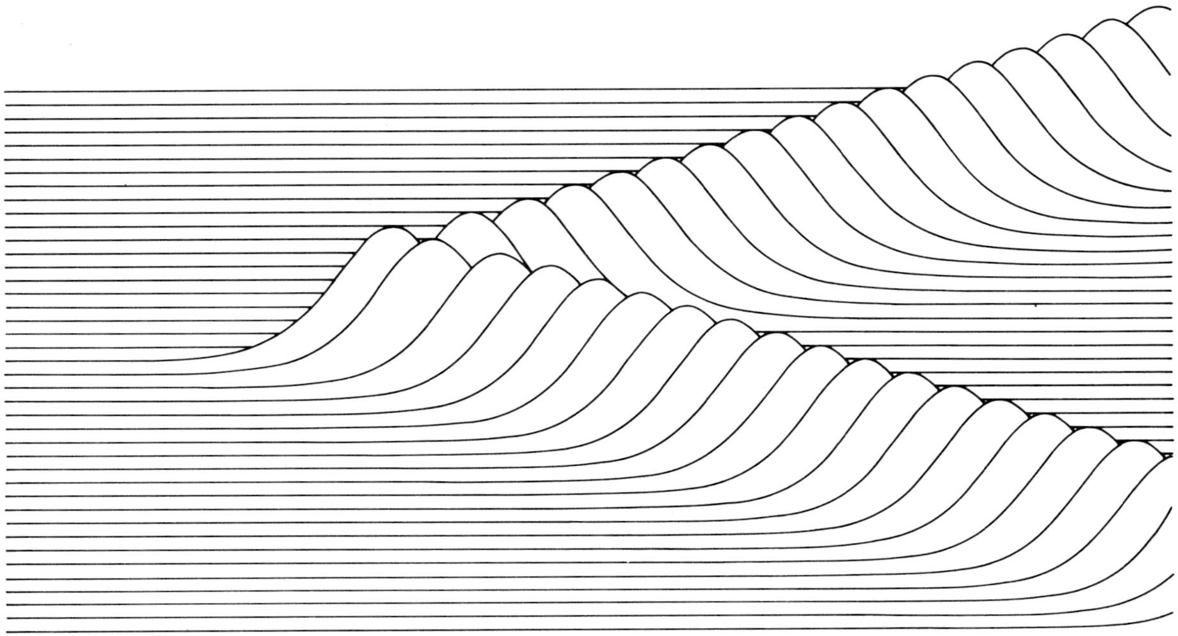

FIGURE 1

Boomeron behaviour. $U_x$ is displayed as a function of x at equally spaced times. U is solution of (2.45).

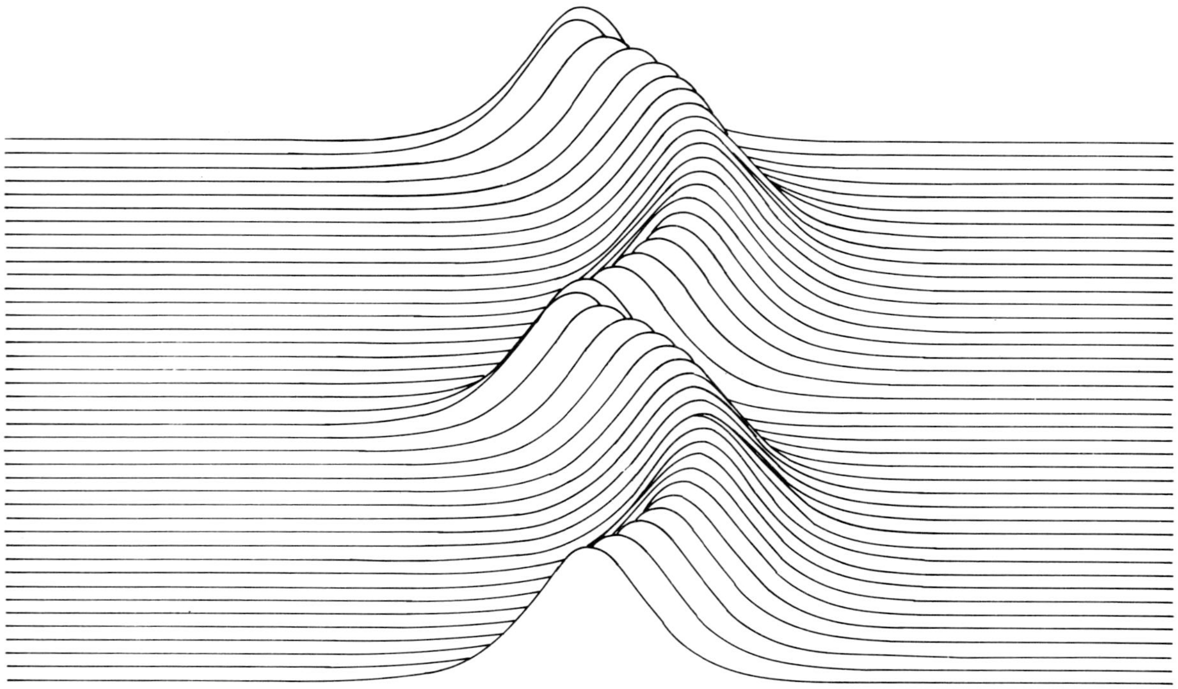

FIGURE 2

Trappon behaviour. $U_x$ and U as in Figure 1.

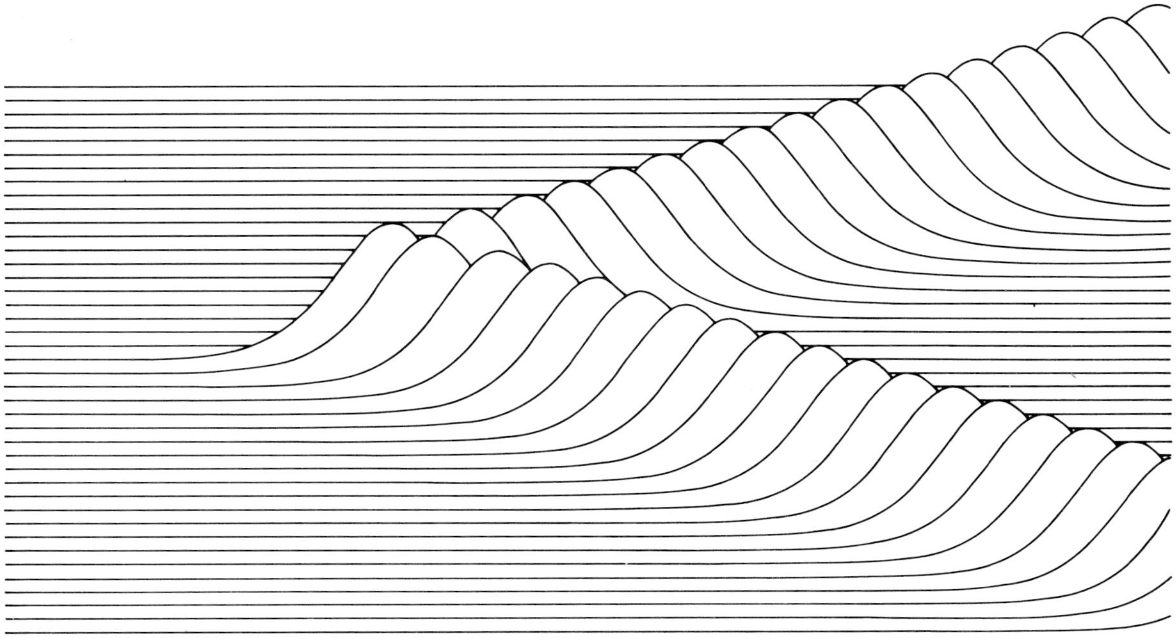

FIGURE 1

Boomeron behaviour. $U_x$ is displayed as a function of x at equally spaced times. U is solution of (2.45).

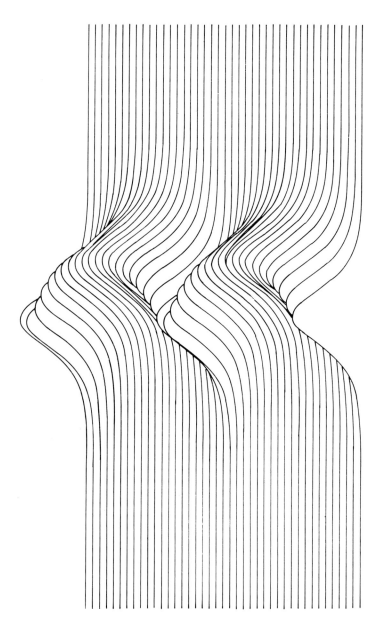

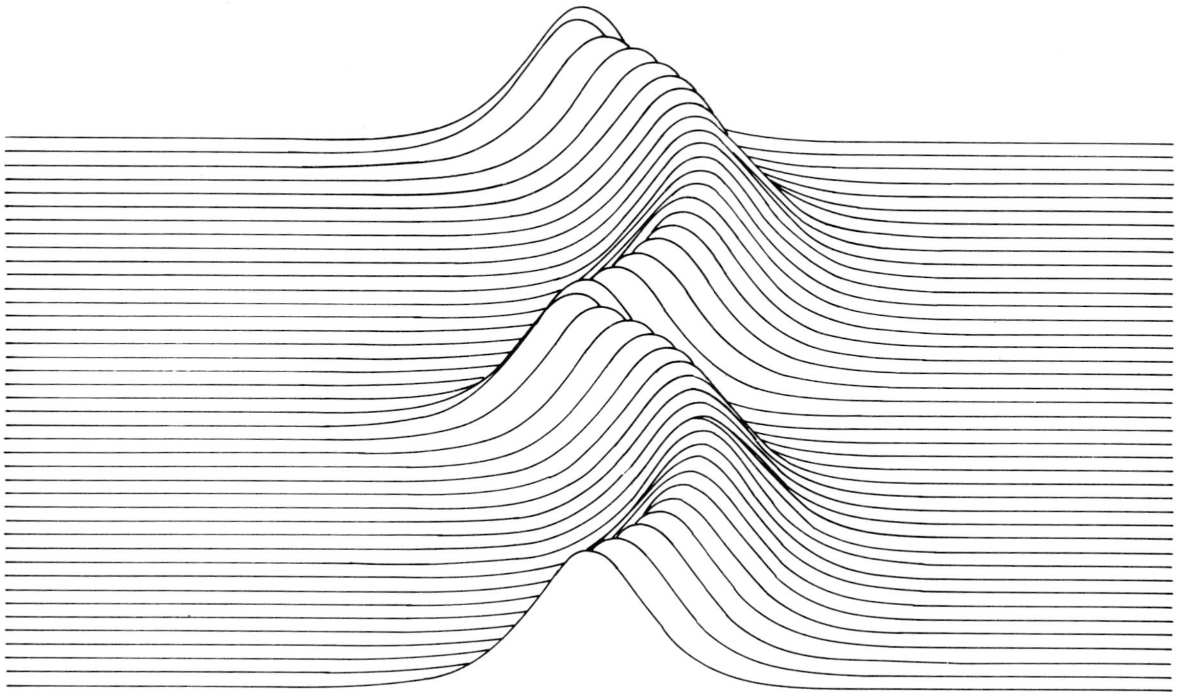

FIGURE 2

Trappon behaviour. $U_x$ and $U$ as in Figure 1.

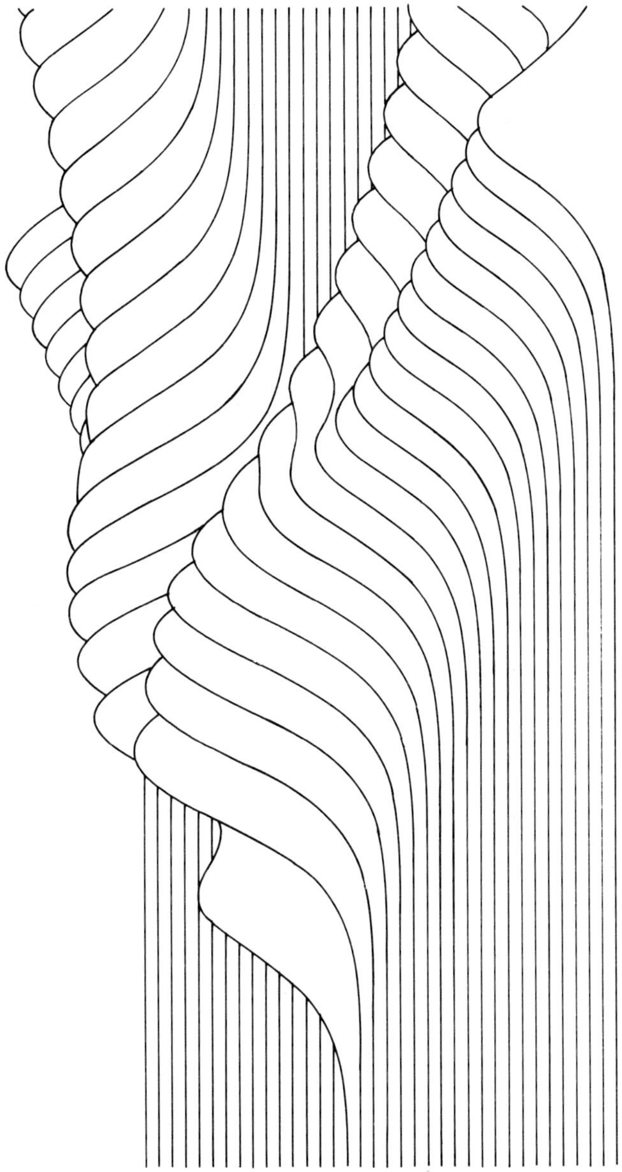

FIGURE 3

Boomeron-Boomeron collision

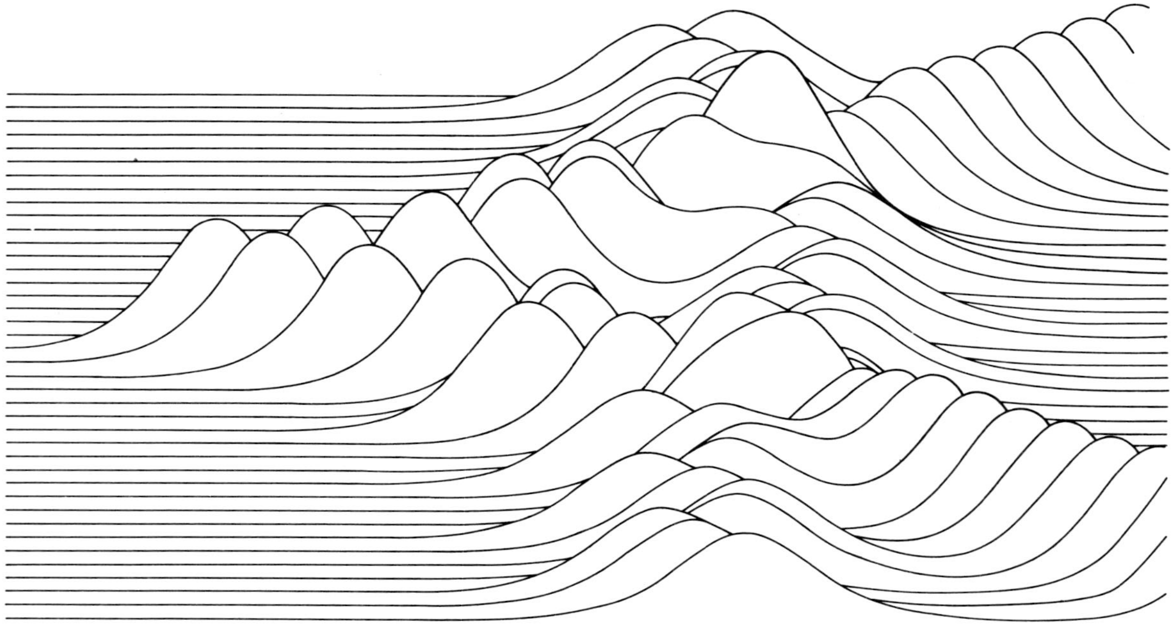

FIGURE 4

Boomeron-Trappon behaviour

All figures have been produced by Dr. J.C. Eilbeck (Heriot-Watt University, Edinburgh, U.K.)

# THE SPECTRAL TRANSFORM AS A TOOL FOR SOLVING NONLINEAR DISCRETE EVOLUTION EQUATIONS

D. Levi
Istituto di Fisica - Universita di Roma, 00185 Roma - Italy
Istituto Nazionale di Fisica Nucleare, Sezione di Roma

Table of Contents

## 1. INTRODUCTION

As it is well known there is an increasing number of problems in natural sciences which are described by nonlinear differential difference equations.

Difference equations arise in many different fields of science, for instance in the study of electrical networks, in statistical problems, in queueing problems, in ecological problems, as computer models for differential equations, etc., that is in all situations in which some sequential relation exist at various discrete values of the independent variable |1|. More specifically, nonlinear discrete evolution equations have been studied as models for vibration of particles in an anharmonic lattice (the Fermi-Pasta-Ulam lattice |2| and the Toda lattice |3|), for treating ladder type electric networks |4|, as a model for wave excitation in plasma |5|, etc.

We can solve the Cauchy problem for many linear discrete evolution equations by a discrete analogue of the Fourier transform |6| which shows itself an equally powerful technique as the continuous one. In the last decades it has been introduced the Spectral Transform method as an extension of the Fourier technique to solve the Cauchy problem for certain classes of nonlinear evolution equations |7|. In the last few years it was recognized that the same kind of ideas, which are at the base of the Spectral Transform method, were applicable to the discrete case as well |8|, thus opening new perspectives in the study of nonlinear discrete evolution problems.

In Section 2 we shall first review the passages necessary to solve linear discrete evolution equations by the discrete Fourier transform; then, starting from the Zakharov-Shabat discretized eigenvalue problem, we shall introduce the Spectral Transform. In Section 3 the correlation between the evolution of the potentials and scattering data will be obtained through the Wronskian technique, obtaining at the same time many properties of the discrete evolution equations, as, for example, the Bäcklund transformations. In Section 4 we shall recover some of the important equations belonging to this class of nonlinear discrete evolution equations and in Section 5 we extend the method to equations with n-dependent coefficients, giving, at the end, the explicit expression of the simplest new equation one obtains in this case.

## 2. THE SPECTRAL TRANSFORM

Let us consider a linear differential-difference equation with constant coefficients

$$u_t(n,t) = - i\omega(E^+,E^-)\, u(n,t) \tag{2.1}$$

where the function u depends on the continuous variable t |9| and on the discrete variable n. The shift operators $E^{\pm}$ are defined by

$$E^{\pm}\phi(n) = \phi(n \pm 1) \tag{2.2}$$

and the function $\omega$ is a real entire or rational function of both its arguments.

Eq.(2.1) can be solved by applying the following integral transform |6|

$$b(z,t) = \sum_{n=-\infty}^{+\infty} u(n,t)\, z^{-n} \tag{2.3a}$$

$$u(n,t) = (1/2\pi i) \oint b(z,t)\, z^{n-1} dz \tag{2.3b}$$

where the loop integral in eq.(2.3b) is around the unit circle in the complex z-plane, if the initial conditions on u are such that b(z,0) is finite. We should like to stress that the integral transform (2.3) is just the analogue of the usual Fourier transform and can be obtained by setting $x = n\Delta$, $z = \exp(ik\Delta)$ and retaining just the lowest term in $\Delta$.

The solution of the Cauchy problem for eq.(2.1) is obtained in 3 steps:

I. starting from the initial data u(n,0) we calculate its transform in the z-space

$$b(z,0) = \sum_{n=-\infty}^{+\infty} u(n,0)\, z^{-n} \tag{2.4a}$$

II. by applying the transformation (2.3) we find the linear differential equation fulfilled by b(z,t):

$$b_t(z,t) = -i\,\omega(z,1/z)\, b(z,t) \tag{2.4b}$$

which yields immediately:

$$b(z,t) = b(z,0)\exp(-i\omega(z,1/z)t) \tag{2.4c}$$

III. The solution of eq.(2.1) is obtained by applying formula (2.3b)·

$$u(n,t) = (1/2\pi i) \sum_{n'=-\infty}^{+\infty} \oint u(n',0)\, z^{n-n'-1} \exp(-i\omega(z,1/z)t)dz \tag{2.4d}$$

We would like to point out that, as in the continuous case, the behaviour of the solution (2.4d) depends essentially on the dispersion function $\omega$ and thus it is much more transparent in the z-space than in configuration space.

The extension of the Fourier transform technique to treat nonlinear differential-difference equations is obtained by introducing a Spectral Transform which, for a given discrete spectral problem,establishes a one-to-one correspondance between the potential u(n) and the spectral data. The firsts who applied to the discrete case the Spectral Transform technique have been Case and Kac |8|, who introduced the spectral transform for a discretized version of the Schrödinger eigenvalue problem

$$u(n)\ \psi(n+1) + u(n-1)\ \psi(n-1) = \lambda\psi(n) \tag{2.5}$$

where $\psi(n)$ is the eigenfunction and $\lambda$ the eigenvalue.

Here we shall instead tersely sketch the Spectral Transform method for a discretized form of the Zakharov-Shabat eigenvalue problem |10|

$$\begin{aligned} \psi_1(n+1) &= \psi_1(n)\ z + Q(n)\ \psi_2(n) \\ \psi_2(n+1) &= \psi_2(n)/z + R(n)\ \psi_1(n) \end{aligned} \tag{2.6}$$

where $\psi_i(n)$ $i = 1,2$ are two fields which take values only for discrete values of their argument n, Q(n) and R(n) are two potentials which go to zero as $|n|$ go to infinity and z is a comples "eigenvalue". Eq.(2.6) is obtained from the Zakharov-Shabat eigenvalue problem |11|

$$\begin{aligned} d\psi_1(x)/dx &= -i\zeta\psi_1(x) + q(x)\ \psi_2(x) \\ d\psi_2(x)/dx &= i\zeta\psi_2(x) + r(x)\ \psi_1(x) \end{aligned} \tag{2.7}$$

by setting

$$\begin{aligned} &\zeta = (i/\Delta)\log z \quad ; \quad q(x) = Q(n)/\Delta \\ &x = n\Delta \qquad\qquad\ \ ; \quad r(x) = R(n)/\Delta \\ &d\psi_i(x)/dx \doteq (\psi_i(n+1) - \psi_i(n))/\Delta \end{aligned} \tag{2.8}$$

By introducing the vector $\underline{\psi}(n)$ of components $\psi_1(n)$ and $\psi_2(n)$ eq.(2.6) reads

$$\underline{\psi}(n+1) = (Z + Q(n)\ \sigma_+ + R(n)\ \sigma_-)\ \underline{\psi}(n) \tag{2.9}$$

with

$$\sigma_\pm = (\sigma_1 \pm i\sigma_2)/2 \quad ; \quad Z = (I + \sigma_3)z/2 + (I - \sigma_3)/2z \tag{2.10}$$

$\sigma_j$ $j = 1,2,3$ being the Pauli matrices and I the identity of rank 2.

Starting from the eigenvalue problem (2.9) for a given set Q(n), R(n) we can define the spectral data

$$S: \{z^{+}_{(k)}(|z^{+}_{(k)}| > 1),\ C^{+}_{(k)}\ ;\quad z^{-}_{(k)}(|z^{-}_{(k)}| < 1),\ C^{-}_{(k)}\ ;\quad (k = 1, \quad , N)\ ;$$

$$\beta^{+}(z)\ (|z| \geq 1),\ \ \beta^{-}(z)\ (|z| \leq 1)\} \tag{2.11}$$

through the asymptotic behaviour of the solution of eq.(2.9), which, for the two independent vector solutions of the scattering process, reads:

$$\Phi(n) = (\underline{\psi}^{1}(n), \underline{\psi}^{2}(n)) \underset{n\to+\infty}{\sim} \begin{pmatrix} z^{n} & ; \beta^{-}(z)z^{n} \\ \beta^{+}(z)z^{-n}; & z^{-n} \end{pmatrix} \tag{2.12a}$$

$$\Phi(n) \underset{n\to-\infty}{\sim} \begin{pmatrix} \alpha^{+}(z)z^{n}\ ; & 0 \\ 0 & ; \alpha^{-}(z)z^{-n} \end{pmatrix} \tag{2.12b}$$

and for the N bound states

$$\underline{\psi}^{\pm}(n) \underset{n\to+\infty}{\sim} C^{\pm}_{(k)}\ \chi^{\mp}\ z^{\mp n}_{(k)} \tag{2.13a}$$

$$\underline{\psi}^{\pm}(n) \underset{n\to-\infty}{\sim} \pm\ \chi^{\pm}\ z^{\pm n}_{(k)} \tag{2.13b}$$

where $\chi^{+} = \begin{pmatrix} 1 \\ 0 \end{pmatrix}$, $\chi^{-} = \begin{pmatrix} 0 \\ 1 \end{pmatrix}$ and $\beta^{+}(z)$ $(\beta^{-}(z))$ is the reflection coefficient, function of z analytic at least outside (inside) the unit circle except for a set of N points $z^{+}_{(k)}$ $(z^{-}_{(k)})$ at which it has a simple pole; if the functions Q(n) and R(n) vanish faster than exponentially as $|n|$ goes to infinity, the constants $C^{\pm}_{(k)}$ are defined as the residues of $\beta^{\pm}(z)$ at $z^{\pm}_{(k)}$.

From a given set of potentials, Q(n) and R(n), we obtain the spectral data S (2.11) through the direct problem (2.9, 2.12, 2.13) and, starting from the spectral data S one recovers in a unique way the potentials by solving the inverse problem through the Gelfand-Levitan-Marchenko discretized matrix equation:

$$K(n,n') + M(n+n') + \sum_{n''=n+1}^{\infty} K(n,n'')\ \sigma_1\ M(n'+n'') = 0 \qquad n' \geq n\ , \tag{2.14}$$

where the kernel of the "integral" equation (2.14) is defined by:

$$M(n) = (I - \sigma_3)\ M^{+}(n)/2 - (I + \sigma_3)\ M^{-}(n)/2 \tag{2.15}$$

$$M^{\pm}(n) = (1/2\pi i) \oint \beta^{\pm}(z)\, z^{n-1} dz \pm \sum_{k=1}^{N} C^{\pm}_{(k)}\, (z^{\pm}_{(k)})^{\mp n-1} \quad , \tag{2.15}$$

and K(n,n') is a matrix of rank 2 whose diagonal elements give back the potentials Q(n) and R(n)

$$Q(n) = -\, K_{22}(n,n+1)\ ;\ R(n) = -\, K_{11}(n,n+1)\ , \tag{2.16}$$

Thus, if Q(n) and R(n) depend on a parameter t, which we shall call time, also the spectral data will depend on t.

Consequently if the evolution of Q(n) and R(n) is governed by a nonlinear discrete evolution equation, while the corresponding evolution of the spectral data is simple, we have a way of finding the solution of the Cauchy problem by an extension of the procedure used in the linear case in eqs. (2.4).

## 3. NONLINEAR DISCRETE EVOLUTION EQUATIONS

One way of obtaining the corresponding evolutions of the potentials and of the spectral data is through the generalized Wronskian relations |12| which are obtained straighforwardly from the eigenvalue problem (2.9) |13|

$$\sum_{n=-\infty}^{+\infty} P(n+1)\Phi'^{T}(n)\ \hat{\eta}(\omega_+(\underline{\Lambda})\ \underline{v}^+(n) + \omega_-(\underline{\Lambda})\ \underline{v}^-(n))\ \Phi(n+1) =$$

$$= \omega_+(z^2) \begin{pmatrix} \beta^+(z) & ; & 1-P(-\infty)\alpha'^+(z)\alpha^-(z) \\ \beta^+(z)\beta'^-(z) & ; & \beta'^-(z) \end{pmatrix} + \omega_-(z^2) \begin{pmatrix} \beta'^+(z) & ; & \beta'^+(z)\beta^-(z) \\ 1-P(-\infty)\alpha^+(z)\alpha'^-(z); & & \beta^-(z) \end{pmatrix}$$

$$+ \begin{pmatrix} 0 & -P(-\infty)\alpha'^+(z)\ \alpha^-(z)\ A^+(z^2) \\ -P(-\infty)\ \alpha^+(z)\ \alpha^-(z)\ A^-(z^2)\ ; & 0 \end{pmatrix} \tag{3.1}$$

with $\underline{v}^+(n) = \begin{pmatrix} R(n) \\ Q'(n) \end{pmatrix}$ , $\underline{v}^-(n) = \begin{pmatrix} R'(n) \\ Q(n) \end{pmatrix}$ , where $\Phi(n)$, respectively $\Phi'(n)$, are two matrix solutions of (2.9) of the same "eigenvalue" z with potentials Q(n) and R(n), respectively Q'(n) and R'(n), and scattering data S, respectively S'; $\Phi'^{T}(n)$ is the transpose of $\Phi'(n)$, P(n) is a scalar defined by the recursive relation

$$P(n) = P(n+1)(1 - Q(n)\ R(n))\ ;\ P(+\infty) = 1$$

$\hat{\eta}$ is an operator such that $\hat{\eta}\begin{pmatrix} a \\ b \end{pmatrix} = \begin{pmatrix} a & 0 \\ 0 & b \end{pmatrix}$, $\omega_{\pm}(x)$ is an arbitrary polinomial function of $x^{\pm 1}$,

$$A^{\pm}(x) = \sum_{k=-\infty}^{+\infty} \{ (\sigma_+ \underline{v}^{\mp}(k), \underline{\Lambda}((\omega_{\pm}(\underline{\Lambda}) - \omega_{\pm}(x))/(\underline{\Lambda} - x))\ \underline{v}^{\pm}(k-1))$$

$$- (\sigma_- \underline{v}^{\mp}(k), \underline{\Lambda}((\omega_{\pm}(\underline{\Lambda}) - \omega_{\pm}(x))/(\underline{\Lambda} - x))\ \underline{v}^{\pm}(k \pm 1)) \}$$

where by (,), here and in the following, we mean the scalar product, and $\underline{\Lambda}$ is a matrix operator of rank 2 such that

$$\underline{\Lambda} \begin{pmatrix} A(n) \\ B(n) \end{pmatrix} = \begin{pmatrix} A(n-1) + R'(n-1)\ a_n (S_n(R(j)\ B(n) - S_n(Q'(j))\ A(n)) \\ B(n+1) + Q(n+1)\ \ a'_n (S_{n+1}(R(j))B(n) - S_{n+1}(Q'(j))\ A(n)) \end{pmatrix.$$

$$\begin{matrix} + R(n)\ (S_n^+(R'(j))B(n) - S_n^-(Q(j))\ A(n)) \\ + Q'(n)(S_{n+1}^+(R'(j))B(n) - S_{n+1}^-(Q(j))A(n)) \end{matrix} \Bigg) \tag{3.2}$$

with

$$a_n = 1 - R(n)\ Q(n) \qquad ; \qquad a'_n = 1 - R'(n)\ Q'(n)$$

$$S_n(y(j))\ x(n) = \sum_{j=n}^{\infty} \left( \prod_{i=n}^{J} \frac{a'_i}{a_i} \right) \frac{X(J)\ Y(J)}{a_j} \quad ; \quad S_n^{\pm}(y(j))\ x(n) = \sum_{j=n}^{\infty} x(j \pm 1)\ y(j)$$

If we set in eq.(3.1)

$$\begin{aligned} Q'(n) &\doteq Q(n) + Q_t(n)\ dt \\ R'(n) &\doteq R(n) + R_t(n)\ dt \end{aligned} \tag{3.3}$$

we obtain, at first order in dt, for $\omega$ (x) = 1

$$\sum_{n=-\infty}^{+\infty} P(n+1)\ \Phi^T(n)\ \hat{n} \begin{pmatrix} -R_t(n) \\ Q_t(n) \end{pmatrix} \Phi(n+1) = \begin{pmatrix} -\beta_t^+ & ; \\ \beta^+(z)\beta_t^-(z) + P(-\infty)\ \alpha^+(z)\ \alpha_t^-(z) & ; \\ -\ (\beta_t^+(z)\ \beta^-(z) + P(-\infty)\alpha_t^+(z)\alpha^-(z) & \\ \beta_t^-(z) & \end{pmatrix} \tag{3.4}$$

$$\beta^+(z)\ \beta^-(z) + P(-\infty)\ \alpha^+(z)\ \alpha^-(z) = 1 \tag{3.5}$$

and for a generic $\omega(x)$

$$\sum_{n=-\infty}^{+\infty} P(n+1)\ \Phi^T(n)\hat{n}\omega(\underline{L}) \begin{pmatrix} R(n) \\ Q(n) \end{pmatrix} \Phi(n+1) = \omega(z^2) \begin{pmatrix} \beta^+(z)\ ; & \beta^+(z)\beta^-(z) \\ \beta^+(z)\ \beta^-(z); & \beta^-(z) \end{pmatrix} \tag{3.6}$$

where $\underline{L}$ is a matrix operator obtained from $\underline{\Lambda}$ by keeping only the first order terms, i.e. by replacing everywhere in eq.(3.2) R' and Q' by R and Q. From eqs.(3.4, 3.6) it follows that, if Q(n) and R(n) evolve according to the nonlinear differential-dif ference equation

$$\begin{pmatrix} R_t(n) \\ -Q_t(n) \end{pmatrix} + \omega(\underline{L}) \begin{pmatrix} R(n) \\ Q(n) \end{pmatrix} = 0 \qquad (3.7)$$

then the evolution of the scattering parameters of the spectral data is simple given by

$$\beta^{\pm}_t(z) \pm \omega(z^2)\ \beta^{\pm}(z) = 0 \qquad (3.8a)$$

$$\alpha^{\pm}_t(z) = 0 \qquad (3.8b)$$

Eqs.(3.8) can be immediately solved to give:

$$\beta^{\pm}(z,t) = \beta^{\pm}(z,0)\ \exp(\mp\omega(z^2)t) \qquad (3.9a)$$

$$\alpha^{\pm}(z,t) = \alpha^{\pm}(z,0) \qquad (3.9b)$$

so that the transmission coefficient is a constant of the motion and the reflection coefficient evolves simple. The evolution of the bound state parameters is simply obtained by calculating the residue of $\beta^{\pm}(z)$ at the poles

$$z^{\pm}_{(k)}(t) = z^{\pm}_{(k)}(0) \qquad (3.10a)$$

$$c^{\pm}_{(k)}(t) = c^{\pm}_{(k)}(0)\ \exp(\mp\,\omega(\,(z^{\pm}_{(k)})^2)t) \qquad (3.10b)$$

thus deducing that the flow is isospectral.

We should like to stress that, as for the Fourier transform, the nonlinear differential-difference equation (3.7) is characterized by the dispersion function $\omega$ which determines the time evolution in the transformed space; it should also be noted the complete analogy between the Spectral Transform method for differential-difference equations and its continuous version |7|: actually this analogy guarantees the possibility of obtaining, also in the discrete case, Bäcklund transformations, the nonlinear superposition principle, the general resolvent formula and an infinity of conserved quantities. From this analogy we can also conclude that also in the discrete case we have two kinds of solutions, soliton solutions and background solutions and any general solution in coordinate space will be a nonlinear superposition of both. For the behaviour of these two kinds of solutions we refer to reference |7| in this same issue.

In the following we just write down the Bäcklund transformations and, from one of them, we shall derive the one soliton solution. Starting from eq.(3.1), if the following equation is fulfilled

$$\omega_+(\underline{\Lambda}) \begin{pmatrix} R(n) \\ Q'(n) \end{pmatrix} + \omega_-(\underline{\Lambda}) \begin{pmatrix} R'(n) \\ Q(n) \end{pmatrix} = 0 \tag{3.11}$$

the right hand side of eq.(3.1) gives:

$$\beta'^{\pm}(z) = -(\omega_{\pm}(z^2)/\omega_{\mp}(z^2))\ \beta^{\pm}(z) \tag{3.12a}$$

$$\alpha'^{\pm}(z) = \omega_{\pm}(z^2)/(\omega_{\pm}(z^2) + A^{\pm}(z^2))\ \alpha^{\pm}(z) \tag{3.12b}$$

showing thus that, if Q(n) and R(n) is a solution of eq.(3.7) so is Q'(n) and R'(n). Eq.(3.11) is thus a Bäcklund transformation. Starting from the solution Q(n)=R(n)=0, we are thus enabled to find the one soliton solution and, from it, to build up the "soliton ladder", that is, step by step, we can construct the N soliton solution. In fact, setting $\omega_+(x) = ax + b$ and $\omega_-(x) = c/x+d$, we obtain from (3.11) the following nonlinear system of difference equations

$$\begin{aligned} Q'(n) &= R'(n)/(R'(n+1)\ R'(n-1)) - 1/R'(n) \\ R'(n) &= Q'(n)/(Q'(n+1)\ Q'(n-1)) - 1/Q'(n) \end{aligned} \tag{3.13}$$

wich determine the one soliton solution as a function of n

$$\begin{aligned} Q'(n) &= C^+\exp(-(p_+(n+1) + p_-(\xi+1)))\ \mathrm{sech}(p_-(n-\xi)) \\ R'(n) &= C^-\exp(p_+n - p_-\xi)\ \mathrm{sech}(p_-(n-\xi)) \\ \xi &= (1/2p_-)\ \log(C^+\ C^-/(\exp(p_+ + p_-)\ \mathrm{sh}^2 p_-)) \\ p_+ &= \log(z^+z^-) \quad ; \quad p_- = \log(z^+/z^-) \end{aligned} \tag{3.14}$$

while the time evolution is recovered by inserting eqs.(3.14) into the evolution equation and is given by eqs.(3.10).

## 4. SPECIAL CASES

In the following pages we want to show some of the interesting equations one can obtain from the class of nonlinear discrete evolution equations given by eq. (3.7).

We consider as a first example the equation one obtains if we set $R(n)=\varepsilon Q^*(n)$, with $\varepsilon = \pm 1$, for Q(n) complex and $\omega(x) = -i(x - 2 + 1/x)$, that is:

$$iQ_t(n) = (Q(n+1) + Q(n-1)(1 + \varepsilon|Q(n)|^2) - 2Q(n) \tag{4.1}$$

This is a differential-difference version of the nonlinear Schrödinger equation |11|, as by setting

$$q(x,t) = Q(n)/\Delta \quad ; \quad x = n\Delta \quad ; \quad T = \Delta^2 t$$
$$Q(n\pm 1) = \Delta q \pm \Delta^2 q_x + (1/2)\,\Delta^3 q_{xx} \pm (1/6)\,\Delta^4 q_{xxx} + O(\Delta^5) \tag{4.2}$$

we get:

$$iq_T = q_{xx} + 2\varepsilon q\,|q|^2 + O(\Delta)$$

A different class of equations obtains in the case $R(n) = \varepsilon Q(n)$ with $Q(n)$ real. By choosing $\omega(x) = x - 1/x + x^2 - 1/x^2$ we obtain:

$$Q_t(n) + (1 - \varepsilon Q^2(n))\,\{Q(n-1) - Q(n+1) + \varepsilon Q(n)(Q^2(N+1) - Q^2(n-1))$$
$$+ Q(n-2)(1 - \varepsilon Q^2(n-1)) - Q(n+2)(1 - \varepsilon Q^2(n+1))\} = 0 \tag{4.3}$$

which results as the appropriate discrete equation for the modified Korteweg-de Vries equation |14| as it reduces to it in the continuous limit (4.2) together with the dispersion relation and its soliton solution (up to terms of order $\Delta^2$) |15|.

Another interesting equation obtains by choosing $\omega(x) = x - 1/x$:

$$Q_t(n) + (1 - \varepsilon Q^2(n))\,(Q(n-1) - Q(n+1)) = 0 \tag{4.4}$$

This equation has been intensively studied as it may represent a model for an electric network as that given in fig.(1), with the inductance and the capacitor depending nonlinearly on the corrent and the potential difference:

$$L = \arctan\,(I(n))/I(n)\ ; \quad C = \arctan\,(V(n))/V(n)$$

in fact, by setting ($\varepsilon = -1$) |16|

$$Q(2n) = -V(n)\ ; \quad Q(2n-1) = -\,I(n)$$

eq.(4.4) reads:

$$dV(n)/dt = (1 + V^2(n))(I(n+1) - I(n))$$
$$dI(n)/dt = (1 + I^2(n))(V(n) - V(n-1)) \tag{4.5}$$

(for $\varepsilon = 1$ it corresponds to the model with $L = \tanh^{-1}(I(n))/I(n)$ $C = \tanh^{-1}(V(n))/V(n)$ which has been studied by Noguchi et al. |17|). By the transformation $N(n) = (Q(n-1) + \sqrt{\varepsilon})(Q(n) - \sqrt{\varepsilon})$ |16| eq.(4.4) reduces to

$$N_t(n) = N(n)\ (N(n+1) - N(n-1)) \tag{4.6}$$

which represent a Volterra model |18| of infinite conflicting species; eq.(4.6) can also describe the evolution of excitations in plasmas |5|. By setting |16|

$$u(n) = -\log(N(2n)\ N(2n-1))$$

for u(n) we get the Toda lattice equations |3|

$$d^2u(n)/dt^2 = 2e^{-u(n)} - e^{-u(n+1)} - e^{-u(n-1)} \tag{4.7}$$

We have to notice that also from eq.(4.4) we can recover, in the continuous limit, the modified Korteweg-de Vries equation, as, by defining |12|

$$Q(n) = \Delta f(X,T)\ ;\quad X = n\Delta + 2\Delta t\ ;\quad T = 3t/\Delta^3$$

we obtain

$$f_T = f_{XXX} + 6\varepsilon f^2 f_X$$

## 5. EXTENSION TO NONLINEAR DIFFERENTIAL-DIFFERENCE EQUATIONS WITH N-DEPENDENT COEFFICIENTS

As is well known, we can apply the Fourier transform method to solve the Cauchy problem for linear discrete evolution equations with n-dependent coefficients at expenses of the simplicity of the evolution equation in the transformed space. In fact, choosing in eq.(2.1) $\omega = (\omega_1(E^+) + \omega_2(E^-))n$, eq.(2.4b) becomes:

$$b_t(z,t) = -i(\omega_1(z) + \omega_2(1/z))z\ b_z(z,t) \tag{5.1}$$

which is a partial differential equation, whose Cauchy problem is solved by the method of characteristics |19|.

It has been shown |20| that also the Spectral Transform method can be extended to treat nonlinear differential-difference equations with n-dependent coefficients by introducing the new Wronskian relation:

$$\sum_{n=-\infty}^{+\infty} P(n+1)\ \Phi^T(n)\ \hat{\eta}\nu(\underline{L})(2n+1) \begin{pmatrix} R(n) \\ Q(n) \end{pmatrix} \Phi(n+1) =$$

$$\nu(z^2)z \begin{pmatrix} \beta_z^+(z) & ;-(\beta^+(z)\ \beta_z^-(z) + P(-\infty)\alpha^+(z)\ \alpha_z^-(z)) \\ \beta_2^+(z)\beta^-(z) + P(-\infty)\ \alpha_z^+(z)\ \alpha^-(z) & ;-\ \beta_z^-(z) \end{pmatrix}$$

$$+ \begin{pmatrix} 0 & ;\ -P(-\infty)\ \alpha^+(z)\ \alpha^-(z)\ B^+(z^2) \\ -P(-\infty)\ \alpha^+(z)\ \alpha^-(z)\ B^-(z^2) & ;\ 0 \end{pmatrix} \tag{5.2}$$

where

$$\nu(z^2) = \nu_+(z^2) + \nu_-(z^2) \tag{5.3}$$

with $\nu_\pm(x)$ a polynomial function of its argument $x^{\pm 1}$, and

$$B^\pm(x) = \sum_{k=-\infty}^{+\infty} \{ (\sigma_+ \underline{v}(k),\ \underline{L}((\nu_\pm(\underline{L}) - \nu_\pm(x))/(\underline{L} - x))\ (2k-1)\ \underline{v}(k-1))$$

$$- (\sigma_- \underline{v}(k),\ \underline{L}((\nu_\pm(\underline{L}) - \nu_\pm(x))/(\underline{L} - x))(2k+1)\ \underline{v}(k+1)) \} \tag{5.4}$$

where $\underline{v}(k) = \begin{pmatrix} R(k) \\ Q(k) \end{pmatrix}$.

Eq.(3.7) thus reads:

$$\begin{pmatrix} R_t(n) \\ -Q_t(n) \end{pmatrix} + \omega(\underline{L}) \begin{pmatrix} R(n) \\ Q(n) \end{pmatrix} + \nu(\underline{L})(2n+1) \begin{pmatrix} R(n) \\ Q(n) \end{pmatrix} = 0 \tag{5.5}$$

and the corresponding linear partial differential equations for the scattering data read:

$$\beta_t^\pm(z) \pm \omega(z^2)\ \beta^\pm(z) + \nu(z^2)\ z\ \beta_z^\pm(z) = 0 \tag{5.6a}$$

$$\alpha_t^\pm(z) + \nu(z^2)\ z\alpha_z^\pm(z) \mp B^\mp(z^2)\ \alpha^\pm(z) = 0 \tag{5.6b}$$

which can be solved by the method of characteristics |19|, thus showing that the flow is no more isospectral and the transmission coefficients are no more constants of the motion.

Work is in progress on this new class of nonlinear differential-difference equations |15, 21|; it is worthwhile to show here just the simplest new equation one obtains in the case of $R(n) = \varepsilon Q(n)$, for $\omega(z) = a(x - 1/x)$ and $\nu(x) = b(x - 1/x)$

$$Q_t(n) + (1 - \varepsilon Q^2(n))(Q(n-1)(2nb + a - b) - Q(n+1)(2nb + a + 3b)) = 0 \tag{5.7}$$

which by the transformation

$$N(n) = K(Q(n-1) + \sqrt{\varepsilon})(Q(n) - \sqrt{\varepsilon}) \tag{5.8}$$

can be reduced to the Volterra type equation

$$N_t(n) = \alpha N(n)(N(n-1)(2n-3+\beta) - N(n+1)\ (2n+3+\beta)) + \gamma N(n) - 2\alpha N^2(n)$$

with $\alpha$, $\beta$, $\gamma$ arbitrary non-zero constants, which is an equation with a dissipative term, a more realistic model for many physical phenomena |5, 22|.

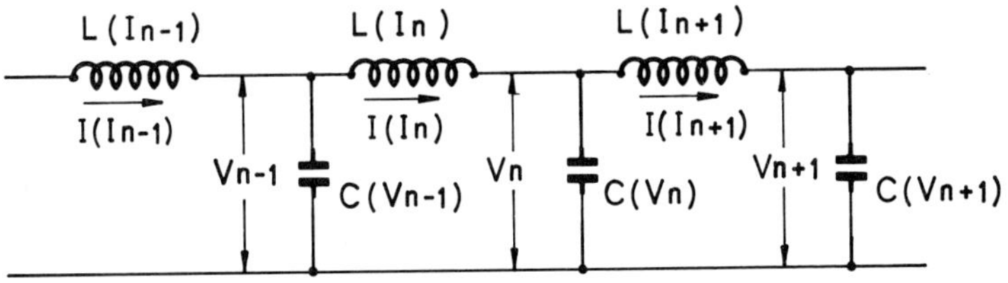


Fig.1. A ladder type nonlinear LC circuit

## REFERENCES

1. H. Levy and F. Lessman. Finite Difference Equations, Macmillan, New York, 1961.

E. Pinney. Ordinary Difference-Differential Equations, Univ. of California Press, New York, 1958.

R. Bellman and K.L. Cooke. Differential-Difference Equations, Academic Press, New York, 1963.

G. Innis. Dynamical analysis in "Soft Science" studies; in defence of difference equations, in Lecture Notes in Biomathematics 2 (S. Levin ed.), Springer Verlag, Berlin 1974, pp. 102-122.

2. E. Fermi, J.R. Pasta and S.M. Ulam. Studies of nonlinear problems, in Collected Papers of E. Fermi, Univ. of Chicago Press, Chicago, 1965, vol.II, pp.977-988.

N.J. Zabusky. A synergetic approach to problems of nonlinear dispersive waves propagation and interaction, in Proc. Symp. on Nonlinear Partial Differential Equations (W.F. Ames, ed.), Academic Press, New York, 1967, pp.223-258.

N.J. Zabusky and M.D. Kruskal. Interaction of "Solitons" in a collisionless plasma and the recurrence of initial states, Phys. Rev. Lett. 15, 240-243 (1965).

3. M. Toda. Vibration of a chain with nonlinear interaction, J. Phys. Soc. Japan 22, 431-436 (1967).

M. Toda. Wave propagation in anharmonic lattice, J. Phys. Soc. Japan 23, 501-506 (1967).

M. Toda. Mechanics and statistical mechanics of nonlinear chains, in Proc. of the International Conference on Statistical Mechanics, Kyoto 1968, Suppl. J. Phys. Soc. Japan 26, 235-237 (1969).

M. Toda. Waves in nonlinear lattice, Suppl. Progr. Theor. Phys. 45, 174-200 (1970).

M. Toda and M. Wadati. A soliton and two solitons in an exponential lattice and related equations, J. Phys. Soc. Japan 34, 18-25 (1973).

R. Hirota. Exact N-soliton solution of a nonlinear lumped network equation, J. Phys. Soc. Japan 35, 286-288 (1973).

M. Hénon. Integrals of the Toda lattice, Phys. Rev. B9, 1921-1923 (1974)

H. Flaschka. The Toda lattice I, existance of integrals, Phys. Rev. B9, 1924-1925 (1974).

H. Flaschka. On the Toda lattice II - Inverse scattering solution, Prog. Theor. Phys. 51, 703-716 (1974).

M. Kac and P. van Moerbeke. On an explicit soluble systems of nonlinear differential equations related to certain Toda lattice, Advan. Math. 16, 160-169 (1975).

M. Kac and P. van Moerbeke. On some periodic Toda lattice, Proc. Nat. Acad. Sci. U.S.A. 72, 1627-1629 (1975).

M. Toda. Studies in a non-linear lattice, Phys. Reports 18C, 1-125 (1975).

E. Date and S. Tanaka. Analogue of the inverse scattering theory for the discrete Hill's equation and exact solutions for the periodic Toda lattice, Prog. Theor. Phys. 55, 457-465 (1976).

M. Toda. Development of the theory of a nonlinear lattice, Suppl. Progr. Theor. Phys. 59, 1-35 (1976).

R. Hirota and J. Satsuma. A variety of nonlinear network equations generated from the Bäcklund transformation for the Toda lattice, Suppl. Prog. Theor. Phys. 59, 64-100 (1976).

K. Sawada and T. Kotera. Toda lattice as an integrable system and the uniqueness of Toda's potential, Suppl. Prog. Theor. Phys. 59, 101-106 (1976).

M. Toda, R. Hirota and J. Satsuma. Chopping phenomena of a nonlinear system, Suppl. Prog. Theor. Phys. 59, 148-161 (1976).

H. Flaschka and D.W. McLaughlin. Canonically conjugate variables for the Korteweg-deVries equation and the Toda lattice with periodic boundary conditions, Prog. Theor. Phys. 55, 438-456 (1976).

B.A.Dubrovin, V.B. Matveev and S.P. Novikov. Nonlinear equations of Korteweg-de Vries type, finite-zone linear operators, and abelian varieties, Russian Math. Surveys 31, 59-146 (1976).

R. Hirota. Nonlinear partial difference equations II - Discrete Toda equation, J. Phys. Soc. Japan 43, 2074-2078 (1977).
R.K. Dodd. Generalized Bäcklund transformation for some nonlinear partial difference equations, J. Phys. A11, 81-91 (1978).

4. R. Hirota and K. Suzuki. Studies on lattice solitons by using electric networks, J. Phys. Soc. Japan 28, 1336-1337 (1970).

R. Hirota. Exact N-soliton solution of nonlinear lumped self-dual network equations, J. Phys. Soc. Japan 35, 289-294 (1973).

5. V.E. Zakharov, S.L. Musher and A.M. Rubenchik. Nonlinear stage of parametric wave excitation in a plasma, Sov. Phys. J.E.P.T. Lett. 19, 151-152 (1974).

6. E.C. Titchmarsh. Theory of Fourier integrals, Oxford Univ. Press Oxford 1937, pp.298-301.

E.C. Titchmarsh. Solutions of some functional equations, J. London Math. Soc. 14, 118-124 (1939).

7. F. Calogero. Spectral Transform and nonlinear evolution equations, on this same issue, pp. 29.

A. Degasperis. Spectral Transform and solvability of nonlinear evolution equations, on this same issue, pp. 35.

8. K.M. Case and M. Kac. A discrete version of the Inverse Scattering problem. J. Math. Phys. 14, 594-603 (1973).

K.M. Case. On discrete Inverse Scattering problem II, J. Math. Phys. 14, 916-920 (1973).

K.M. Case and S.C. Chiu. The discrete version of the Marchenko equations in the Inverse Scattering problem. J. Math. Phys. 14, 1643-1647 (1973).

K.M. Case. The discrete Inverse Scattering problem in one dimension, J. Math. Phys. 15, 143-146 (1974).

K.M. Case. Fredholm determinants and multiple solitons, J. Math. Phys. 17, 1703-1706 (1976).

9. Here, and allways in the following, subscripts indicate partial differentiation: $u_t = \partial u/\partial t$, etc.

10. M.J. Ablowitz and J.F. Ladik. Nonlinear differential-difference equations, J. Math. Phys. 16, 598-603 (1975).

M.J. Ablowitz and J.F. Ladik. Nonlinear differential-difference equations and Fourier analysis, J. Math. Phys. 17, 1011-1018 (1976).

M.J. Ablowitz and J.F. Ladik. A nonlinear difference scheme and Inverse Scattering, Stud. Appl. Math. 55, 213-229 (1976).

M.J. Ablowitz. Nonlinear evolution equations-continuous and discrete, Siam Review 19, 663-684 (1977).

M.J. Ablowitz. Lectures on the Inverse Scattering transform, Stud. Appl. Math. 58, 17-94 (1978).

11. V. Zakharov and A.B. Shabat. Exact theory of two-dimensional self-focusing and one-dimensional self-modulation of waves in nonlinear media, Soviet. Phys. J.E.P.T. 34, 62-69 (1972).

12. S.C. Chiu and J.F. Ladik. Generating exactly soluble nonlinear discrete evolution equations by a generalized Wronskian technique, J. Math. Phys. 18, 690-700 (1977).

13. For another method of obtaining the nonlinear discrete evolution equations, see the works of ref. 8

14. M. Wadati. The exact solution of the Modified Korteweg-deVries equation, J.Phys. Soc. Japan 32, 1681 (1972).

15. M. Bruschi, D. Levi and O. Ragnisco. Spectral Transform approach to a discrete version of the Modified Korteweg-de Vries equation with x-dependent coefficients, sent to the Il Nuovo Cimento for pubblication.

16. M. Wadati. Transformation theories for nonlinear discrete systems, Suppl. Prog. Theor. Phys. 59, 36-63 (1976).

17. A. Noguchi, H. Watanabe and K. Sakai. Shock waves and hole type solitons of nonlinear selfdual network equation, J. Phys. Soc. Japan 43, 1441-1446 (1977).

18. V. Volterra. Lecons sur la théorie mathématique de la lutte pour la vie, Gautier-Villars, Paris 1931.

R. Hirota and J. Satsuma. N-solitons solutions of nonlinear network equations describing a Volterra system, J. Phys. Soc. Japan 40, 891-900 (1976).

S. Fujii, F. Kako and N. Mugibayashi. Inverse method applied to the solution of nonlinear network equations describing a Volterra system, J. Phys. Soc. Japan 42, 335-340 (1977).

J.F. Ladik and S.C. Chiu. Solution of nonlinear network equations by the Inverse Scattering method, J. Math. Phys. 18, 701-704 (1977).

M. Wadati and M. Watanabe. Conservation laws of a Volterra system and nonlinear self-dual network equations, Prog. Theor. Phys. 57, 808-811 (1977).

19. G.F. Carrier and C.E. Pearson. Partial differential equations, theory and technique, Academic Press, New York 1976.

20. D. Levi and O. Ragnisco. Extension of the Spectral Transform method for solving nonlinear differential-difference equations, Lett. Nuovo Cimento (1978), 22 691-696.

21. M. Bruschi, D. Levi and O. Ragnisco. Spectral Transform approach to a discrete version of the nonlinear Schrödinger equation with x-dependent coefficients, in preparation.

22. J.C. Fernandez and G. Reinisch. Collapse of a Volterra soliton into a weak monotone shock wave, Physica A (1978) in press.

# HYPERBOLIC BALANCE LAWS IN CONTINUUM PHYSICS

C.M. Dafermos
Brown University, Providence RI, USA

Table of Contents

## 1. INTRODUCTION

After half a century of decline, there has been, in the last three decades, a revival of interest in continuum physics, due partly to the development of new materials of technological importance (high polymers, rubbers, liquid crystals) and partly to the realization that modern Geometry and Analysis can be employed in order to recast the classical theories into a more rational formulation and solve old problems that had been abandoned as unsolvable. In the first five sections of these lecture notes, I attempt to present a bird's-eye view of the structure of continuum physical theories today.

In rough terms, a continuum physical theory consists of a collection of balance laws and a set of constitutive relations. The former identifies the framework of the theory (e.g., continuum mechanics, continuum thermomechanics, continuum electrodynamics, etc.) while the latter characterizes the nature of the continuous medium (e.g., elastic solid, viscous fluid, etc.). In Section 3 we discuss the general form of a balance law and write down, as an illustration, the balance laws of continuum thermomechanics.

The major success of modern continuum physics is the development of a rational theory of constitutive relations, founded on the principles of frame indifference, material symmetry and an appropriate interpretation of the second law of thermodynamics. No attempt will be made here to present this theory in its generality; however, the basic ideas will be discussed in the framework of the very simple concrete example of hyperelasticity (Section 5). In the remainder of the lecture notes, we study certain aspects of the theory of nonlinear hyperbolic systems that result from the balance laws of continuum physics. This is a branch of the theory of partial differential equations that is being studied vigorously at the present time. We discuss here the main difficulties of the subject, namely nonexistence of smooth solutions (breaking of waves, development of shock waves) and nonuniqueness of weak solutions and we show how ideas originating in continuum physics (the second law of thermodynamics) can be employed in order to single out the physically admissible solution.

I wish to thank the organizers of the symposium and in particular Antonio F.Rañada and Luis and Luciana Vazquez for their hospitality.

## 2. KINEMATICS OF BODIES

In continuum physics, the mathematical model of a <u>body</u> is a manifold characterized by a <u>reference configuration</u>, that is an open subset B of the reference space $R^3$. The typical point $\underset{\sim}{X} \in B$ will be called a <u>material particule</u>. In the applications, the standard method of constructing a reference configuration of a moving body is to identify "molecules" with the point in space that they happen to occupy at a certain fixed time instant. In general, however, a reference configuration need not be an actual configuration of the body but only an abstract three-dimensional "picture" of it. As a matter of fact it is convenient to assume throughout that the reference space is not necessarily the physical space but just a copy of it.

A <u>configuration</u> of the body B is a diffeomorphism $\underset{\sim}{x} = \underset{\sim}{x}(\underset{\sim}{X})$ from B to the physical space $R^3$. A <u>motion</u> of B is a family $\underset{\sim}{x} = \underset{\sim}{x}(\underset{\sim}{X},t)$ of configurations, parametrized by time t. Thus, for fixed $(\underset{\sim}{X},t)$, $\underset{\sim}{x}(\underset{\sim}{X},t)$ will be the position of particle $\underset{\sim}{X}$ at time t; for fixed t, the map $\underset{\sim}{x}(.,t): B \to R^3$ is the configuration of the body at time t; finally, for fixed $\underset{\sim}{X} \in B$, the curve $\underset{\sim}{x}(\underset{\sim}{X},.)$ is the trajectory of particle $\underset{\sim}{X}$.

In continuum physics one seeks to determine the evolution in time of various fields of physical quantities, such as temperature, electric field, etc. defined over the moving body. Since every configuration is a diffeomorphism of the reference configuration, the field quantities can be represented equally well as functions of $(\underset{\sim}{X},t)$ (referential description)or $(\underset{\sim}{x},t)$ (spatial description). Both representations are useful. This induces a nasty notation problem since three different symbols are needed for each field quantity, namely one for the value of the quantity, one for its referential description and one for its spatial description. For example, in the case of temperature,

$$\theta = \Phi(\underset{\sim}{X},t) = \phi(\underset{\sim}{x},t) \tag{2.1}$$

In order to circumvent this cumbersome notation one generally identifies functions with their values, using, for example, $\theta$ to denote the value of temperature as well as the functions $\Phi$ and $\phi$ in (2.1). To avoid confusion in the representation of partial derivatives one makes the following convention: Cartesian components of $\underset{\sim}{X}$, in reference space, will be designated with Greek superscripts, such as $X^\alpha$, $X^\beta$, $X^\gamma$, etc., while Cartesian components of $\underset{\sim}{x}$, in physical space, will be designated with lower case Latin superscripts, e.g., $x^i$, $x^j$, $x^k$, etc. In the referential representation of a field, the partial derivatives $\frac{\partial}{\partial X^\alpha}$ and $\frac{\partial}{\partial t}$ will be denoted by $,\alpha$ and a dot (.), respectively, while in the spatial representation $\frac{\partial}{\partial x^i}$ and $\frac{\partial}{\partial t}$ will be denoted by ,i and $\frac{\partial}{\partial t}$. Thus, in the example (2.1) of temperature,

$$\frac{\partial\Phi}{\partial X^\alpha} = \theta_{,\alpha} \quad , \quad \frac{\partial\Phi}{\partial t} = \dot{\theta} \quad , \quad \frac{\partial\phi}{\partial x^i} = \theta_{,i} \quad , \quad \frac{\partial\phi}{\partial t} = \frac{\partial\theta}{\partial t}$$

We also adopt throughout the usual repeated index summation convention. It turns out that the above notation is very efficient and the reader rapidly becomes familiar with it.

The basic kinematic fields associated with a motion $\underset{\sim}{x} = \underset{\sim}{x}(\underset{\sim}{X},t)$ of the body, are <u>velocity</u> $\underset{\sim}{v} = \dot{\underset{\sim}{x}}$ and <u>deformation gradient</u> $\underset{\sim}{F} = \nabla_X \underset{\sim}{x}$ (components $F^i_\alpha = x^i_{,\alpha}$). Since every configuration is a diffeomorphism of the reference configuration we must have $\det F \neq 0$.

Chain rule induces obvious relations between the various partial derivatives. For example, in the case of temperature,

$$\dot{\theta} = \frac{\partial\theta}{\partial t} + v^i\theta_{,i} \quad , \quad \theta_{,\alpha} = F^i_\alpha\, \theta_{,i}$$

## 3. BALANCE LAWS

The referential description of a balance law for a moving body B is an equation of the form

$$\frac{d}{dt}\int_\Omega G\, d\underset{\sim}{X} = \oint_{\partial\Omega} P dS + \int_\Omega H\, d\underset{\sim}{X} \tag{3.1}$$

that holds for every smooth subset $\Omega$ of B with boundary $\partial\Omega$. Equation (3.1) states that the quantity whose density is G is conserved in the sense that the time rate of change of the amount contained in $\Omega$ is balanced by the rate of flux through $\partial\Omega$ and the rate of production in $\Omega$. In (3.1) G and H are fields, i.e., $G = G(\underset{\sim}{X},t)$, $H = H(\underset{\sim}{X},t)$. However, P does not only depend on $(\underset{\sim}{X},t)$ but also upon characteristics of $\partial\Omega$ in the vicinity of $\underset{\sim}{X}$. The crucial idea was contributed by Cauchy who postulated (for the typical case of balance of momentum) that P may depend solely on the orientation of $\partial\Omega$ at $\underset{\sim}{X}$, that is, $P = P(\underset{\sim}{X},t,\underset{\sim}{N})$ where $\underset{\sim}{N}$ is the unit normal on $\partial\Omega$ at the point $\underset{\sim}{X}$. Cauchy then proceeded to prove that the dependence upon $\underset{\sim}{N}$ is necessarily linear, i.e.,

$$P = P^\alpha(\underset{\sim}{X},t)\, N_\alpha \tag{3.2}$$

The reason of this surprising property is that, as $\Omega$ shrinks about a point, the volume of $\Omega$ goes to zero faster than the area of $\partial\Omega$ and, as a result, the flux term $\int_\Omega P dS$ must balance locally in itself. For a formal proof, using the celebrated "Cauchy tetrahedron argument" see, e.g. |1|.

In view of (3.2), (3.1) takes the form

$$\frac{d}{dt}\int_\Omega G\, d\underset{\sim}{X} = \oint_{\partial\Omega} P^\alpha N_\alpha dS + \int_\Omega H\, d\underset{\sim}{X} \tag{3.3}$$

Applying the Gauss-Green theorem, one easily shows that (3.3) is equivalent to the field equation

$$\dot{G} = P^{\alpha}_{,\alpha} + H \tag{3.4}$$

The cumbersome integral equation (3.1) has thus been replaced by a differential equation. It is this event that makes continuum physics workable!

The balance law (3.3) also admits the spatial description

$$\frac{d}{dt}\int_{\omega(t)} g d\underset{\sim}{x} = \oint_{\partial\omega(t)} p^i n_i ds + \int_{\omega(t)} h d\underset{\sim}{x} \tag{3.5}$$

where $\omega(t)$ is the configuration of $\Omega$ at time t, $\underset{\sim}{n}$ is the unit normal on $\partial\omega(t)$ and

$$g = (\det \underset{\sim}{F})^{-1} G \ , \qquad h = (\det \underset{\sim}{F})^{-1} H \ , \qquad p^i = (\det \underset{\sim}{F})^{-1} F^i_{\alpha} P^{\alpha} \tag{3.6}$$

Since $\omega(t)$ varies with t,

$$\frac{d}{dt}\int_{\omega(t)} g d\underset{\sim}{x} = \int_{\omega(t)} \frac{\partial g}{\partial t} d\underset{\sim}{x} + \oint_{\partial\omega(t)} g v^i n_i ds$$

so that (3.5) yields the field equation

$$\frac{\partial g}{\partial t} + (g v^i)_{,i} = p^i_{,i} + h \tag{3.7}$$

Continuum mechanics is characterized by the balance laws of mass, linear momentum and moment of momentum. For the referential description of the balance law of mass, one takes the reference mass density $\rho_0$ for G and zero for $P^{\alpha}$ and H. Thus (3.4) yields $\dot{\rho}_0 = 0$, i.e., the reference density is independent of t. For the spatial description of the some balance law, one takes the spatial mass density $\rho = (\det \underset{\sim}{F})^{-1} \rho_0$ for g and zero for $p^i$ and h, so that (3.7) gives

$$\frac{\partial \rho}{\partial t} + (\rho v^i)_{,i} = 0 \tag{3.8}$$

In the referential description of the balance law of linear momentum G is $\rho_0 \underset{\sim}{v}$, $P^{\alpha}$ is $T^{\alpha}_i$, the Piola-Kirchhoff stress tensor, and H is $\rho_0 \underset{\sim}{b}$, the body force. Thus the field equation is

$$\rho_0 \dot{v}_i = T^{\alpha}_{i,\alpha} + \rho_0 b_i \tag{3.9}$$

In the corresponding spatial description, g is $\rho \underset{\sim}{v}$, $p^i$ is $\tau^i_j$, the Cauchy stress tensor, and h is $\rho \underset{\sim}{b}$. The field equation reduced with the help of (3.8), takes the form

$$\rho \dot{v}_i = \tau^j_{i,j} + \rho b_i \tag{3.10}$$

The field equations of the balance law of moment of momentum can be reduced considerably with the help of the previous balance laws and end up in algebraic form:

$$F^i_\alpha \, T^\alpha_j = F^j_\alpha \, T^\alpha_i , \tag{3.11}$$

$$\tau^i_j = \tau^j_i \tag{3.12}$$

Thus the Cauchy stress tensor is symmetric.

In continuum thermomechanics, the balance law of energy should be added to the above list. Its referential description is obtained by inserting $\rho_o \varepsilon + \frac{1}{2} \rho_o \, |\underset{\sim}{v}|^2$ for G, $T^\alpha_i \, v^i + Q^\alpha$ for $P^\alpha$ and $\rho_o b_i v^i + \rho_o r$ for H, where $\varepsilon$ is the internal energy, $\underset{\sim}{Q}$ is the heat flux vector and r is the energy supply. For the spatial description we take $\rho \varepsilon + \frac{1}{2} \rho |\underset{\sim}{v}|^2$ for g, $\tau^i_j v^j + q^i$ for $p^i$ and $\rho b_i v^i + \rho r$ for h. The corresponding field equations, reduced with the help of the balance laws of mass and momentum, are

$$\rho_o \dot{\varepsilon} = T^\alpha_i v^i_{,\alpha} + Q^\alpha_{,\alpha} + \rho_o r \quad , \tag{3.13}$$

$$\rho \dot{\varepsilon} = \tau^j_i v^i_{,j} + q^j_{,j} + \rho r \tag{3.14}$$

The above list of balance laws should be supplemented with the second law of thermodynamics, expressed by the <u>Clausius-Duhem inequality</u>

$$\rho_o \dot{\eta} - \left(\frac{Q^\alpha}{\theta}\right)_{,\alpha} - \rho_o \frac{r}{\theta} \geq 0 \quad , \tag{3.15}$$

$$\rho \dot{\eta} - \left(\frac{q^i}{\theta}\right)_{,i} - \rho \frac{r}{\theta} \geq 0 \quad , \tag{3.16}$$

in referential and spatial description, respectively, where $\eta$ is the specif entropy and $\theta$ is the absolute temperature.

In continuum electrodynamics we also have the balance laws of electric charge and magnetic flux but we shall not write them down since the examples considered thus far are probably sufficient form to illuminate the structure of balance laws in continuum physics.

## 4. CONSTITUTIVE RELATIONS

Constitutive relations relate the various fields that enter in the balance laws and characterize the nature of the continuous medium (e.g. viscous fluid, plasma, elastic dielectric, etc.). For example, constitutive relations determine, in continuum mechanics, stress from the kinematical quantities and, in continuum thermomechanics, stress, internal energy, heat flux and entropy (or temperature) from the kinematical quantitites and temperature (or entropy).

Constitutive relations must be compatible with the balance law of moment of momentum, (3.11), (3.12), as well as the Clausius-Duhem inequality (3.15), (3.16). They must also comply with the principle of frame indifference which relates the fields corresponding to any two motions differing by a rigid rotation. Finally, constitutive relations must reflect the type of material symmetry (e.g. isotropy) with which the medium happens to be endowed.

In the following section we show, in the framework of a simple example, how constitutive relations are reduced in order to comply with the above requirements. There is an extensive theory of constitutive relations for which the reader is referred to |2|.

## 5. HYPERELASTICITY

In the framework of continuum mechanics, a material is hyperelastic if the Piola-Kirchhoff stress is determined from deformation gradient by a constitutive relation of the form

$$T_i^\alpha = \frac{\partial W(\underset{\sim}{F})}{\partial F_\alpha^i} \tag{5.1}$$

The strain energy function $W(\underset{\sim}{F})$ is interpreted as the density of the (potential) mechanical energy stored in the body by deformation.

In the context of hyperelasticity, the principle of frame indifference states, that strain energy should be unaffected by rigid rotations, that is the strain energies of any two configurations $\underset{\sim}{x} = \underset{\sim}{x}(\underset{\sim}{X})$ and $\underset{\sim}{x} = \underset{\sim}{x}^*(\underset{\sim}{X}) = \underset{\sim}{Q}\underset{\sim}{x}(\underset{\sim}{X})$, with $\underset{\sim}{Q}^T\underset{\sim}{Q} = I$, are the same. Since $\underset{\sim}{F}^* = \nabla_X \underset{\sim}{x}^* = \underset{\sim}{Q}\,\nabla_X \underset{\sim}{x} = \underset{\sim}{Q}\underset{\sim}{F}$, the above requirement means

$$W(\underset{\sim}{F}) = W(\underset{\sim}{Q}\,\underset{\sim}{F}), \quad \text{for all } \underset{\sim}{F} \text{ nonsingular, } \underset{\sim}{Q} \text{ orthogonal} \tag{5.2}$$

By the polar decomposition theorem we may write $\underset{\sim}{F} = \underset{\sim}{R}\,\underset{\sim}{U}$, where the rotation tensor $\underset{\sim}{R}$ is orthogonal and the right stretch tensor $\underset{\sim}{U} = \sqrt{\underset{\sim}{F}^T\underset{\sim}{F}}$ is symmetric and positive definite. Applying (5.2) with $\underset{\sim}{Q} = \underset{\sim}{R}^T$ we deduce

$$W = W(\underset{\sim}{U}) \; , \tag{5.3}$$

that is, frame indifference precludes dependence of strain energy upon the rotation tensor and allows only dependence on the right stretch tensor. It is more convenient to visualize W as a function of the right Cauchy-Green strain tensor $\underset{\sim}{C} = \underset{\sim}{U}^2 = \underset{\sim}{F}^T \underset{\sim}{F}$, i.e.,

$$W = W(\underset{\sim}{C}) \tag{5.4}$$

It is easy to check that the Piola-Kirchhofff stress, determined by (5.1),with W given by (5.4), satisfies automatically the balance law of moment of momentum (3.11). This happy coincidence, however, is peculiar to hyperelasticity since in more general theories the balance law of moment of momentum imposes on constitutive relations additional restrictions to those dictated by frame indifference.

We now consider material symmetry in hyperelasticity. If $W(\underset{\sim}{C})$ and $W^*(\underset{\sim}{C}^*)$ are the strain energy functions of the same material relative to two different reference configurations B and $B^*$, related via the diffeomorphism $\underset{\sim}{X} = \underset{\sim}{X}(\underset{\sim}{X}^*)$, we must have

$$W(\underset{\sim}{C}) = W^*(\underset{\sim}{C}^*) \; , \quad \text{whenever} \quad \underset{\sim}{C}^* = \underset{\sim}{H}^T \underset{\sim}{C} \, \underset{\sim}{H} \, , \tag{5.5}$$

where $\underset{\sim}{H} = \partial\underset{\sim}{X}/\partial\underset{\sim}{X}^*$. When the material is endowed with symmetry, certain reference configurations $B^*$ will be indistinguishable from B in the sense that $W^*(.) = W(.)$. In order to identify those $B^*$ and on account of (5.5) we consider the class $\Gamma$ of unimodular matrices $\underset{\sim}{H}$ (we need $|\det \underset{\sim}{H}| = 1$ to avoid changes in reference density) with the property

$$W(\underset{\sim}{C}) = W(\underset{\sim}{H}^T \underset{\sim}{C}\,\underset{\sim}{H}) \quad , \quad \text{for all symmetric positive definite } \underset{\sim}{C} \; . \tag{5.6}$$

It is easily verified that $\Gamma$ forms a subgroup of the unimodular (or special linear) group $SL_3$. Thus $\Gamma$ is called the isotropy group of the material, relative to the reference configuration B. In practice one observes the material symmetry of a given material determines the isotropy group $\Gamma$ and then seeks the form of the strain energy function $W(\underset{\sim}{C})$ which is compatible with that $\Gamma$. In order to illustrate this procedure we will now discuss two examples.

A hyperelastic material with maximal symmetry, i.e., $\Gamma = SL_3$, is called a hyperelastic fluid. In this case (5.6) will be satisfied for every unimodular matrix $\underset{\sim}{H}$ so we may select, in particular, $\underset{\sim}{H} = \underset{\sim}{H}^T = (\det \underset{\sim}{C})^{1/6} \, \underset{\sim}{C}^{-1/2}$, in which case $\underset{\sim}{H}^T \underset{\sim}{C}\,\underset{\sim}{H} = (\det \underset{\sim}{C})^{1/3} \, \underset{\sim}{1}$. Since $\det \underset{\sim}{C} = (\det \underset{\sim}{F})^2 = \rho_0^2/\rho^2$, we deduce that a hyperelastic material is a fluid if and only if the strain energy is a function of density,

$W = W(\rho)$. A simple calculation yields

$$\tau^i_j = -\frac{\rho^2}{\rho_0}\frac{dW(\rho)}{d\rho}\delta^i_j , \tag{5.7}$$

that is, the Cauchy stress in a hyperelastic fluid is a hydrostatic pressure depending on density. Thus the hyperelastic fluid is just the ideal fluid of classical physics.

As a second example we consider the isotropic hyperelastic solid, characterized by the condition that the isotropy group $\Gamma$ is the full orthogonal group. Thus properties of the material are the same in every direction in reference space. To determine the form os strain energy we diagonalize the symmetric matrix $\underset{\sim}{C}$, $\underset{\sim}{C} = \underset{\sim}{Q}\,\underset{\sim}{L}\,\underset{\sim}{Q}^T$, where $\underset{\sim}{Q}^T\underset{\sim}{Q} = I$, and then insert $\underset{\sim}{H} = \underset{\sim}{Q}$ in (5.6). It follows that $W = W(\underset{\sim}{L})$ is a symmetric function of the eigenvalues of the Cauchy-Green strain tensor $\underset{\sim}{C}$. Equivalently, one may visualize the strain energy as a function $W = W(I, II, III)$ of the principal invariants $I = \text{tr}\,\underset{\sim}{C}$, $II = \frac{1}{2}\left|(\text{tr}\,\underset{\sim}{C})^2 - \text{tr}\,\underset{\sim}{C}^2\right|$, $III = \det \underset{\sim}{C}$ of $\underset{\sim}{C}$.

A typical problem in hyperelasticity theory is to determine the motion $\underset{\sim}{x} = \underset{\sim}{x}(\underset{\sim}{X},t)$ of a body B by solving the field equations (3.9) for given reference density $\rho_0(\underset{\sim}{X})$ and body force $\underset{\sim}{b}(\underset{\sim}{X},t)$ and with prescribed initial conditions $\underset{\sim}{x}(\underset{\sim}{X},0)$, $\underset{\sim}{v}(\underset{\sim}{X},0)$ and boundary conditions (assigning, for example, forces on the boundary). This is a formidable problem and very little is known concerning even gross qualitative behaviour, such as existence, uniqueness and stability of solutions. The discussion of hyperbolic systems in the following sections will shed some light on the difficulties of the problem.

## 6. HYPERBOLIC BALANCE LAWS

In the remainder of the notes we shall be looking into the theory of balance laws from a somewhat more formalistic point of view. We callect all balance laws of our theory into a single vector-valued balance law

$$\underset{\sim}{V} = \underset{\sim}{P}^\alpha_{,\alpha} + \underset{\sim}{H} , \tag{6.1}$$

where $\underset{\sim}{V}$, $\underset{\sim}{P}^\alpha$ and $\underset{\sim}{H}$ are n-dimensional vector fields whose Cartesian components will be denoted by capital Latin subscripts, such as $V_A$, $P^\alpha_A$, $H_B$, etc. The state of each particle will be expressed by an n-dimensional state vector $\underset{\sim}{U}$. The fields $\underset{\sim}{V}$, $\underset{\sim}{P}^\alpha$ and $\underset{\sim}{H}$ will be determined from $\underset{\sim}{U}$ through smooth constitutive type relations

$$\underset{\sim}{V} = \underset{\sim}{V}(\underset{\sim}{U}) , \quad \underset{\sim}{P}^\alpha = \underset{\sim}{P}^\alpha(\underset{\sim}{U}) , \quad \underset{\sim}{H} = \underset{\sim}{H}(\underset{\sim}{U}, \underset{\sim}{X}, t) \tag{6.2}$$

(as before, we identify functions with their values). We assume that $\underset{\sim}{U} \to \underset{\sim}{V}(\underset{\sim}{U})$ is a diffeomorphism so that (6.1) becomes a system of equations for determining $\underset{\sim}{U}(\underset{\sim}{X}, t)$.

For a homogeneous ($\rho_0$ = const.) hyperelastic material the state vector $\underset{\sim}{U} = (F^1_1, F^1_2, \ldots, F^3_3, v^1, v^2, v^3)$ is 12-dimensional and the list of balance laws reads

$$\dot{F}^i_\alpha = v^i_{,\alpha} \qquad i,\alpha = 1,2,3$$
$$\rho_0 \dot{v}_i = T^\alpha_{i,\alpha} + \rho_0 b_i \qquad i = 1,2,3. \tag{6.3}$$

We may rewrite (6.1) in the form

$$\frac{\partial V_A}{\partial U_B} \dot{U}_B = \frac{\partial P_A}{\partial U_B} U_{B,\alpha} + H_A \tag{6.4}$$

System (6.4) will be called <u>hyperbolic</u> if for any fixed $\underset{\sim}{U} \in R^n$ and every 3-dimensional unit vector $\underset{\sim}{\xi}$ the eigenvalue problem

$$(\lambda \nabla \underset{\sim}{V} + \xi_\alpha \nabla \underset{\sim}{P}^\alpha) \underset{\sim}{E} = \underset{\sim}{0} \tag{6.5}$$

has real eigenvalues and n linearly independent eigenvectors. The physical interpretation of hyperbolicity is understood when one considers weak waves, that is solutions of (6.4) which are $C^m$-smooth, $m \geq 0$, but whose derivatives of order $m+1$ experience jump discontinuities across a propagating surface (wave). One easily shows that if the direction of propagation is $\underset{\sim}{\xi}$, then the speed of propagation and amplitude of the wave are, respectively, an eigenvalue and a corresponding eigenvector of (6.5).

In particular, (6.3) is hyperbolic if and only if the strain energy function $W(\underset{\sim}{F})$ satisfies the Hadamard (or strong ellipticity, or rank-one convexity) condition

$$\frac{\partial^2 W(\underset{\sim}{F})}{\partial F^i_\alpha \partial F^j_\beta} \zeta^i \zeta^j \xi_\alpha \xi_\beta > 0 \tag{6.6}$$

for all nonzero 3-vectors $\underset{\sim}{\xi}$ and $\underset{\sim}{\zeta}$.

## 7. <u>SHOCK WAVES</u>

From the point of view of analysis, the difficulty with nonlinear hyperbolic systems is that in general there are no globally defined smooth solutions. This is due to the property that wave speeds (the eigenvalues of (6.5)) are not constant but depend on the solution itself so that waves catch up with one another, are amplified, and eventually break. One is then looking for solutions in the class of functions that are smooth except on a family of smooth propagating surfaces (shock waves) across which they experience jump discontinuities. A function $\underset{\sim}{U}(\underset{\sim}{X},t)$ in this class is a weak solution of (6.1) (that is a solution of the balance law in integral form) if it satisfies (6.1) at every point of smoothness while across surfaces of discontinuity the Rankine-Hugoniot jump conditions

$$s\,[\underset{\sim}{V}] = -\,\xi_\alpha\,[\underset{\sim}{P}^\alpha] \tag{7.1}$$

hold, where $\underset{\sim}{\xi}$ is the direction and s is the speed of propagation of the wave and a bracket denotes the jump of the enclosed quantity across the surface of discontinuity.

The class of piecewise smooth functions is not sufficiently broad to encompass the solutions of (6.1) under all (even $C^\infty$-smooth!) initial data. It has been conjectured that solutions of (6.1) are generically piecewise smooth but so far this has only been established in the case of a very simple model equation. It seems that solutions of (6.1) should be sought in the class of functions of bounded variation in the sense of Tonelli-Cesari, i.e., the class of locally bounded measurable functions whose distributional derivatives are locally Borel measures. Functions of bounded variation are endowed with a geometric structure that resembles the structure of piecewise smooth functions so that, in the framework of these solutions, one may talk, in an appropriately generalized sense, about shock waves, the Rankine-Hugoniot conditions, etc.

## 8. ENTROPY

One of the remarkable features of the theory of discontinuous solutions of (6.1) is nonuniqueness. In continuum thermomechanics, in order to single out the physically admissible solutions, one makes an appeal to the Clausius-Duhem inequality (3.15), (3.16). To extend this idea to the present, more abstract, set up, we assume that one may append to our balance law (6.1) an "entropy" inequality

$$\dot{\eta} \geq S^\alpha_{,\alpha} + R\,, \tag{8.1}$$

where the entropy $\eta$, entropy flux $\underset{\sim}{S}$ and entropy production R are determined by $\underset{\sim}{U}$ via constitutive type relations

$$\eta = \eta(\underset{\sim}{U})\,,\quad S^\alpha = S^\alpha(\underset{\sim}{U})\,,\quad R = R(\underset{\sim}{U},\underset{\sim}{X},t) \tag{8.2}$$

A solution of (6.1) will be called admissible if it satisfies (8.1). It is standard practice in continuum physics to require that all smooth solutions be admissible so we also impose this condition here, assuming that

$$\frac{\partial \eta}{\partial U_B} \dot{U}_B = \frac{\partial S^\alpha}{\partial U_B} U_{B,\alpha} + R \tag{8.3}$$

for every smooth solution $\underset{\sim}{U}(\underset{\sim}{X},t)$ of (6.4). This assumption induces the existence of an n-vector valued integrating factor

$$\underset{\sim}{L} = \underset{\sim}{L}(\underset{\sim}{U}) \tag{8.4}$$

with the property

$$\frac{\partial \eta}{\partial U_N} = L^A \frac{\partial V_A}{\partial U_N} , \tag{8.5}$$

$$\frac{\partial S^\alpha}{\partial U_N} = L^B \frac{\partial P^\alpha_B}{\partial U_N} , \tag{8.6}$$

$$R = L^A H_A \tag{8.7}$$

In the example (6.3) of homogeneous hyperelasticity,

$$\eta = - \frac{1}{2} \rho_o |\underset{\sim}{v}|^2 - W(\underset{\sim}{F}) , \tag{8.8}$$

$$S^\alpha = - v^i T^\alpha_i , \tag{8.9}$$

$$R = - \rho_o v^i b_i , \tag{8.10}$$

$$\underset{\sim}{L} = - (T^1_1, T^1_2, \ldots, T^3_3, v^1, v^2, v^3) , \tag{8.11}$$

so that the physical interpretation of "entropy" is minus mechanical energy.

Our objective is to investigate uniqueness and stability of admissible solutions by using the methodology of D. Perna |3|.

Since $\underset{\sim}{U} \to \underset{\sim}{V}(\underset{\sim}{U})$ is a diffeomorphism, we may visualize $\underset{\sim}{U}, \underset{\sim}{P}^\alpha, \eta, S^\alpha, \underset{\sim}{L}$ as functions of $\underset{\sim}{V}$ and $\underset{\sim}{H}, R$ as functions of $(\underset{\sim}{V}, \underset{\sim}{X}, t)$. Then (8.5), (8.6) take the form

$$\frac{\partial \eta}{\partial V_M} = L^M , \tag{8.12}$$

$$\frac{\partial S^\alpha}{\partial V_M} = L^B \frac{\partial P^\alpha_B}{\partial V_M} \tag{8.13}$$

We now define

$$A(\underset{\sim}{V},\bar{\underset{\sim}{V}}) = \eta(\bar{\underset{\sim}{V}}) - \eta(\underset{\sim}{V}) - \underset{\sim}{L}^{A}(\underset{\sim}{V})(\bar{V}_A - V_A) \ , \tag{8.14}$$

$$D^{\alpha}(\underset{\sim}{V},\bar{\underset{\sim}{V}}) = S^{\alpha}(\bar{\underset{\sim}{V}}) - S^{\alpha}(\underset{\sim}{V}) - \underset{\sim}{L}^{A}(\underset{\sim}{V})\left[P^{\alpha}_{A}(\bar{\underset{\sim}{V}}) - P^{\alpha}_{A}(\underset{\sim}{V})\right] . \tag{8.15}$$

Assume that $\underset{\sim}{U}(\underset{\sim}{X},t)$ is a smooth (i.e. Lipschitz continuous) solution of (6.1) and $\underset{\sim}{U}(\underset{\sim}{X},t)$ is any admissible solution of (6.1), of bounded variation. Using (6.1), (8.1) and (8.3) we deduce

$$\dot{A} - D^{\alpha}_{,\alpha} \geq \bar{R} - R - L^{A}(\bar{H}A - H_A) - \dot{L}^{A}(\bar{V}_A - V_A) + L^{A}_{,\alpha}(\bar{P}^{\alpha}_{A} - P^{\alpha}_{A}) \tag{8.16}$$

By virtue of (6.1)

$$\dot{L}^{A} = \frac{\partial L^{A}}{\partial V_B}\dot{V}_B = \frac{\partial L^{A}}{\partial V_B}\frac{\partial P^{\alpha}_{B}}{\partial V_M}V_{M,\alpha} + \frac{\partial L^{A}}{\partial V_B}H_B \tag{8.17}$$

However, on account of (8.12) and (8.13)

$$\frac{\partial L^{A}}{\partial V_B} = \frac{\partial L^{B}}{\partial V_A} \ , \tag{8.18}$$

$$\frac{\partial L^{B}}{\partial V_A}\frac{\partial P^{\alpha}_{B}}{\partial V_M} = \frac{\partial L^{B}}{\partial V_M}\frac{\partial P^{\alpha}_{B}}{\partial V_A} \tag{8.19}$$

Thus (8.17) yields

$$\dot{L}^{A} = \frac{\partial L^{B}}{\partial V_M}\frac{\partial P^{\alpha}_{B}}{\partial V_A}V_{M,\alpha} + \frac{\partial L^{B}}{\partial V_A}H_B = \frac{\partial P^{\alpha}_{B}}{\partial V_A}L^{B}_{,\alpha} + \frac{\partial L^{B}}{\partial V_A}H_B \tag{8.20}$$

Furthermore, from (8.7),

$$\bar{R} - R - L^{A}(\bar{H}_A - H_A) = H_A(\bar{L}^{A} - L^{A}) + (\bar{H}_A - H_A)(\bar{L}^{A} - L^{A}) \tag{8.21}$$

Combining (8.16), (8.20), (8.21) and interchanging appropriately the dummy summation indices A and B we obtain

$$\dot{A} - D^{\alpha}_{,\alpha} \geq \{\bar{P}^{\alpha}_{A} - P^{\alpha}_{A} - \frac{\partial P^{\alpha}_{A}}{\partial V_B}(\bar{V}_B - V_B)\}L^{A}_{,\alpha} + \{\bar{L}^{A} - L^{A} - \frac{\partial L^{A}}{\partial V_B}(\bar{V}_B - V_B)\}H_A$$
$$+ (\bar{H}_A - H_A)(\bar{L}^{A} - L^{A}). \tag{8.22}$$

The crucial observation is that A and $D^{\alpha}$, as given by (8.14) and (8.15), as well as the right-hand side of (8.22) are of quadratic order in $\bar{V} - V$. Suppose that the flow of $D$ through the boundary of the body vanishes (in the hyperelasticit, case this condition will be satisfied when, for example, the forces excerted upon the boundary are equal or when the boundary moves with the same velocity in the two solutions). Assume, further, that $A(V,\bar{V})$ is negative definite. Then, integrating (8.22) over the body and using Gronwall's inequality, one easily arrives at an estimate of the form

$$\int_B |\bar{V}(X,t) - V(X,t)|^2 dX \leq K e^{kt} \int_B |\bar{V}(X,0) - V(X,0)|^2 \, dX , \qquad (8.23)$$

where K and k are positive constants depending on the solution $V(X,t)$. In particular, when $\bar{V}(X,0) = V(X,0)$, $X \in B$, the two solutions coincide. We have thus shown that, under the negative definiteness assumption on A, whenever a smooth solution of (6.1) exists then there is no other admissible solution within the broader class of functions of bounded variation, that satisfies the same initial and boundary conditions.

In view of (8.14) and (8.12), A will be globally negative definite if and only if $\eta$ is a uniformly concave function of $V$. In the applications to continuum physics, this condition is sometimes satisfied and sometimes not. For example, in the hyperelasticity case, $\eta$, as expressed by (8.8), will be concave if and only if the strain energy W is a convex function of deformation gradient $F$. It can be shown, however, that, due to the representation (5.4), dictated, as we have seen, by the principle of frame indifference, W can n ever be a globally convex function of $F$. Even so, it turns cut that W is convex on a large region in phase space, so that we at least have local uniqueness and stability of smooth solutions with range in that region of convexity.

In the most common case where no smooth solution of (6.1) exist the entropy admissibility criterion is no longer sufficiently powerful to single out the physically relevant solution by ruling out all extraneous ones. Stronger admissibility criteria that have been proposed include Liu's shock admissibility condition |4|, the viscosity criterion and the entropy rate criterion |5|. The entropy rate criterion requires that admissible solutions maximize at every point the expression $\dot{\eta} - S^{\alpha}_{,\alpha} - R$, i.e., whenever there are many solutions that satisfy the entropy inequality (8.1) then the admissible is the one that maximizes the rate of entropy production. The compatibility of the various admissibility criteria has been established in special examples but the general problem is still open.

## REFERENCES

|1| Truesdell, C.A. and Toupin, R.A., The Classical Field Theories. Handbuch der Physik (S. Flügge, Ed.), vol.III/1. Berlin: Springer-Verlag 1970.

|2| Truesdell, C.A. and Noll, W., The Non-Linear Field Theories of Mechanics. Handbuch der Physik (S. Flügge, Ed.),vol.III/3. Berlin: Springer-Verlag 1965.

|3| Diperna, R.J. Uniqueness of solutions to hyperbolic conservation laws (to appear).

|4| Liu, T.P., The entropy condition and the admissibility of shocks. J. Math. Anal. Appl. 53, 78-88 (1976).

|5| Dafermos, C.M., The entropy rate admissibility criterion in thermoelasticity. Rend. Accad. Naz. Lincei, Ser. VIII, 57, 113-119 (1974).

# MATHEMATICAL ASPECTS OF CLASSICAL NONLINEAR FIELD EQUATIONS

W. Strauss
Brown University, Providence R.I.

Table of Contens

## INTRODUCTION

Most of the classical nonlinear field equations are far too difficult for mathematical analysis at this time. However in recent years there have been some important advances for the simples equations. For this reason I will emphasize in these lectures the model equations:

$$i \frac{\partial u}{\partial t} - \Delta u + F(u) = 0 \qquad \text{(NLS)}$$

the nonlinear Schrödinger equation, and

$$\frac{\partial^2 u}{\partial t^2} - \Delta u + m^2 u + F(u) = 0 \qquad \text{(NLKG)}$$

the nonlinear Klein-Gordon equation. Here $\Delta = \partial^2/\partial x_1^2 + \ldots \partial^2/\partial x_n^2$ is the Laplacian, $x = (x_1,\ldots,x_n)$ is in n-space $\mathbb{R}^n$, and $F(0) = F'(0) = 0$.

It is usually easy to solve such equations for a short time (or "locally in time" as we say) but we are interested in global questions. What general properties does the complete time evolution of the system have and in particular what is its asymptotic behavior?. Specifically nonlinear phenomena only become evident in the long-time behavior. These include under various circunstances: (i) the development of shock waves (not for NLS or NLKG but for the equations discussed by Prof. Dafermos in his lectures), (ii) blow up of solutions in a finite time, (iii) the existence of solitons, a kind of nonlinear generalization of bound states, and (iv) relativistic scattering phenomena. It must be emphasized that solutions of nonlinear equations cannot be combined linearly to obtain new solutions and in particular they cannot be constructed by Fourier analysis.

The plan of these lecture notes is as follows. In chapter I we discuss NLS, its application to the theory of lasers, and its bound states. In chapter II we discuss NLKG and other relativistic wave equations and the general question of the existence of solutions. In chapter III we discuss the conservation laws which follow from the invariance properties of the equations and apply them in particular to the Yang-Mills equations. In chapter IV we discuss scattering theory mainly in the context of NLKG. The references listed are not meant to be complete or even representative of the topics covered.

# I. THE NONLINEAR SCHRÖDINGER EQUATION

## I.1 Trapping and focusing of laser beams

Consider the electromagnetic wave equation

$$\frac{1}{c^2}\frac{\partial^2}{\partial t^2}(\varepsilon\vec{E}) - \Delta\vec{E} = 0 \tag{1}$$

Assume a linearly polarized wave ($\vec{E}$ parallel to a fixed unit vector $\vec{e}$) which is monochromatic with frequency $\omega$ and which propagates along the z-axis. Thus

$$\vec{E} = u(x,y,z)\exp(i(kz - \omega t))\,\vec{e}$$

and (1) reduces to

$$2ik\frac{\partial u}{\partial z} - \Delta u + (k^2 - \varepsilon\omega^2/c^2)\,u = 0 \tag{2}$$

The high intensity of a laser beam can produce significant local changes in the density of the medium and hence in the dielectric constant $\varepsilon$.

Chiao, Garmire and Townes |1| assume the simple nonlinear dependence $\varepsilon = \varepsilon_0 + \varepsilon_2\,|\vec{E}|^2$. They show how the resulting nonlinear term may give rise to an electromagnetic beam which produces its own waveguide and propagates without spreading. This phenomenon is called "self-trapping". It corresponds to a solution of (2) which is independent of z:

$$-\frac{\partial^2 u}{\partial x^2} - \frac{\partial^2 u}{\partial y^2} + (k^2 - \varepsilon_0\frac{\omega^2}{c^2})\,u - \varepsilon_2\frac{\omega^2}{c^2}|u|^2 u = 0 \tag{3}$$

Kelley and Talanov |2| show how a nonlinear dependence of $\varepsilon$ can produce a build-up in the intensity of part of the beam as a function of z. This phenomenon is called "self-focusing". It corresponds to a solution of (2) in which the intensity $|u|^2$ blows up at a certain value of z. If the "paraxial" approximation $|u_{zz}| \ll |ku_z|$ is used and we choose $k = \omega\sqrt{\varepsilon_0}/c$, equation (2) reduces to

$$2ik\frac{\partial u}{\partial z} - \frac{\partial^2 u}{\partial x^2} - \frac{\partial^2 y}{\partial y^2} - k\frac{\varepsilon_2}{\varepsilon_0}|u|^2 u = 0 \tag{4}$$

Under appropriate assumptions about the nature of the absorbing medium, the beam can be spread out, distorted or bent instead focused. These phenomena are called "thermal blooming". If we take into account the effects of a steady transverse wind along the y-axis, then $\varepsilon - \varepsilon_0$ depends on the beam intensity $|u|^2$ ups upstream. In the simplest case we have

$$\varepsilon - \varepsilon_0 = \text{const} \int_{-\infty}^{y} |u|^2 dt \tag{5}$$

A similar equation describes the long-time behavior of one-dimensional hydromagnetic waves in a plasma:

$$i \frac{\partial u}{\partial t} - c \frac{\partial^2 u}{\partial x^2} - (a|u|^2 + b)u = 0 \tag{6}$$

where a,b,c are constants. Such an equation also occurs in the Ginzburg-Landau theory of superconductivity.

## I.2 Existence and blow-up

We consider the problem

$$i \frac{\partial u}{\partial t} - \Delta u + \lambda |u|^{p-1} u = 0 \tag{7}$$

$$u(x,0) = \phi(x) \quad , \quad x \in \mathbb{R}^n$$

where $\phi$ is a nice function ($\phi \in S$ if not stated otherwise), $p > 1$ and $\lambda$ is real. The two standard conservation laws are obtained by multiplying (7) by $\bar{u}$ and taking imaginary parts, and by multiplying by $\bar{u}_t$ and taking real parts. They are

$$\int |u|^2 \, dx = \text{constant} \qquad \text{and}$$

$$E = \int \{ \frac{1}{2} |\nabla u|^2 + \frac{\lambda}{p+1} |u|^{p+1} \} dx = \text{constant} .$$

They would lead us to believe that solutions ought to exist globally and be stable if $\lambda > 0$, but that if $\lambda < 0$ instability may be possible.

Theorem 1.1 There exists a unique solution in some time interval $\{x \in \mathbb{R}^n, |t| < t_1\}$. It is as smooth as the singularity at $u = 0$ of the function $|u|^{p-1} u$ allows; if p is an odd integer it is $C^\infty$.

This local existence result follows easily by standard methods (See Section II.4).

Theorem 1.2 Let $\lambda < 0$ and $p \geq 1 + 4/n$. Let $\phi$ satisfy the condition

$$E = \int \{ \frac{1}{2} |\nabla\phi|^2 dx + \frac{\lambda}{p+1} |\phi|^{p+1} \} dx < 0$$

Then no smooth solution of (7) can exist for all t.

In case $p = 3$, $n = 2$, the equation reduces to (4) and Theorem 2 provides a precise initial condition for the self-focusing of an electromagnetic beam. The opposite situation is described in the next theorem.

Theorem 1.3 If $\lambda < 0$ assume $p < 1 + 4/n$. If $\lambda > 0$ and $n > 2$ assume $p < 1 + 4/(n-2)$. If $\phi \in H^1(\mathbb{R}^n)$, there exists a unique solution of (7) for all t which is a bounded continuous function of t with values in $H^1(\mathbb{R}^n)$. The regularity statement of Theorem 1 is valid for all t.

In particular, equation (6) has global smooth solutions.

In view of Theorems 2 and 3, we call $\lambda > 0$ and $\lambda < 0$, $p < 1 + 4/n$ the stable cases and we call $\lambda < 0$, $p \geq 1 + 4/n$ the unstable case.

Theorem 1.4 Let $\lambda > 0$ and $p < \infty$. If $\phi \in H^1(\mathbb{R}^n)$, there exists a solution which is a bounded, weakly continuous function of t with values in $H^1(\mathbb{R}^n)$.

This theorem is proved below. We do not know whether this weak solution is unique or smooth.

Theorem 1.5 For any $\lambda$, there exists a unique smooth solution of

$$i \frac{\partial u}{\partial t} - \frac{\partial^2 y}{\partial x^2} - \frac{\partial^2 y}{\partial y^2} + \lambda u \int_{-\infty}^{y} |u(x,y',t)|^2 dy' = 0$$

with $u(x,y,0) = \phi(x,y)$. (See |3|).

Proof of Theorem 1.2 Write $F(u) = \lambda|u|^{p-1} u$ and $G(u) = \lambda|u|^{p+1}/(p+1)$. Multiply equation (7) by $2r\bar{u}_r + n\bar{u}$ where $r = |x|$, and take real parts:

$$\frac{d}{dt} \operatorname{Im} \int r u_r \bar{u}\, dx = -2 \int |\nabla u|^2 dx + n \int \left| 2G(u) - \bar{u}F(u) \right| dx$$

$$= -4E + \int \left| (2n+4) G(u) - n\bar{u}F(u) \right| dx$$

$$\geq -4E \quad \text{since} \quad p \geq 1 + 4/n .$$

Now multiply (7) by $r^2\bar{u}$ and take imaginary parts:

$$\frac{d}{dt}\int r^2|u|^2 dx = -\,4\ \mathrm{Im}\int r u_r \bar{u}\ dx\ .$$

Hence

$$\frac{d^2}{dt^2}\int r^2|u|^2 dx \leq 16E < 0\ .$$

This is nonsense because the last integral is positive.

Proof of Theorem 1.3 We merely sketch the ingredients. (i) If $p < 1 + 4/(n-2)$, Sobolev's theorem states that $H^1 \quad L^{p+1}$. (ii) The evolution operator $U_0(t) : \phi \to u(t)$ for the free (linear) Schrödinger equation satisfies the estimate

$$\|U_0(t)\phi\|_q \leq t^{-\delta}\|\phi\|_{q'}$$

for $2 \leq q \leq \infty$, $1/q + 1/q' = 1$, $\delta = n/2 - n/q$, as can be obtained by interpolation from the cases $q = 2$ and $q = \infty$. If $q = p + 1$, then $0 < \delta < 1$ so that $t^{-\delta}$ is integrable at $t = 0$. (iii) We use the conservation laws. If $\lambda > 0$, both terms in E are non-negative and hence bounded for all t. If $\lambda < 0$, Sobolev's inequality states that

$$\int |u|^{p+1} dx \leq c\,\{\int |u|^2 dx\}^{\alpha}\,\{\int |\nabla u|^2 dx\}^{\beta}$$

where $2\alpha = (2-n)p + 2 + n$ and $\beta = n(p-1)/4$. If $p < 1 + 4/n$, we have $\alpha < 0$ and $\beta < 1$. Hence the conservation laws provide a bound in $H^1$ and in $L^{p+1}$. (iv) To prove the uniqueness, let u and v be two solutions. By (ii)

$$\|u(t) - v(t)\|_{p+1} \leq c\int_0^t (t-s)^{-\delta}\,\|\,|u|^{p-1}u - |v|^{p-1}v\|_{(p+1)/p}(s)\ ds$$

$$\leq c\int_0^t (t-s)^{-\delta}\|u - v\|_{p+1}(s)\ ds$$

since u and v are bounded in $L^{p+1}$, by (iii). This implies $u = v$ because $\delta < 1$. (v) The existence and regularity are proved by combining (i)-(iii) with Theorem 1.

Proof of Theorem 1.4 Let $g(s) = \lambda s^{p+1}/(p+1)$ for $s \geq 0$. We approximate $g(s)$ by a sequence of smooth functions $g_N(s)$ such that $g_N(s) = \mathrm{const}\ s^2$ for s large and $0 \leq g_N(s) \leq g(s)$, $g_N(s) \to g(s)$ for all s. Let $F(u) = \lambda|u|^{p-1}u$ and $F_N(u) = g_N'(|u|)u/|u|$. Let $u_N(x,t)$ be the unique solution of the problem

$$iu_{Nt} - \Delta u_N + F_N(u_N) = 0\ ,\quad u_N(x,0) = \phi(x) \tag{8}$$

It exists because $F_N(u)$ is linear for large u so that an easy variation of Theorem 3 is applicable. Then we have the obvious a priori bounds

$$\int |u_N|^2 dx = \int |\phi|^2 \, dx \qquad \text{and}$$

$$\int \{ \tfrac{1}{2} |\nabla u_N|^2 + g_N(|u_N|) \} \, dx \leq \int \{ \tfrac{1}{2} |\nabla \phi|^2 + g(|\phi|) \} \, dx$$

By compactness, there is a subsequence (still called $u_N$) which converges to some u weakly in $H^1$ and almost everywhere. Hence $g_N(|u_N|) \to g(|u|)$ a.e. and $F_N(u_N) \to F_N(u)$ locally in $L^1$, by Theorem 1.1 of |4|. We may pass to the limit in each term in (8). The other properties of u follow easily as in |4|.

## 1.3 Bound States

If we substitute a standing wave $u = \phi(x) \exp(-i\omega t)$ into NLS, we get the elliptic equation

$$\omega\phi - \Delta\phi + F(\phi) = 0 \tag{9}$$

provided that $\arg F(u) = \arg u$. This is the same equation as (3) for a particular F. Into NLS we can also substitute the more general expression

$$u = \phi(x - ct) \exp \left[ ig(x - bt) \right]$$

where $\phi$ and g are real scalar functions and c and b are constant vectors. This is a solution of NLS provided $\phi$ satisfies (9) and $g(x) = -c.x/2 + g_0$, $g_0$ = constant and $4\omega = c.(c-2b) > 0$.

We define a generalized bound state as a solution of NLS which has finite energy (is square-integrable) for each t but whose amplitude $\max_x |u(x,t)|$ does not go to zero as $|t| \to \infty$. Then we are that every solution of (9) leads to a whole family of generalized bound states of NLS depending on several parameters. Here are some examples of equations of the form (9). We take $n = 3$. By scaling ($x \to cx$ and $\phi \to c\phi$) we can adjust the values of various constants; for instance we can always assume $\omega = 1$ in (9). We only consider solutions which vanish at infinity and which are not identically zero.

The equation $-\Delta\phi + \phi - |\phi|^{q-1}\,\phi = 0$ possesses an infinite sequence of distinct solutions if $1 < q < 5$, but no solutions (of finite energy) if $q \geq 5$. These solution decay exponentially as $|x| \to \infty$. The equation $-\Delta\phi - \phi + \lambda\sin\phi = 0$ possesses at least one (non-trivial) solution if $\lambda > 1$. The equation

$$-\Delta\phi + \phi + |\phi|^{p-1}\phi - \lambda|\phi|^{q-1}\phi = 0$$

has at least one solution if $1 < q < \max(p,5)$ and $\lambda > \lambda_*$ where $\lambda_* = \lambda_*(p,q)$ is explicitly computable. The equation $-\Delta\phi - |\phi|^{q-1} = 0$ possesses a solution only if $q = 5$ in which case $\phi(x) = c(c^4 + |x|^2)^{-1/2}$ is one. The equation $-\Delta\phi + |\phi|^{p-1}\phi - |\phi|^{q-1}\phi=0$ has a solution if $5 < q < p$ and if $p < q < 5$. For another explicit example, see the lectures of Prof. Bialynicki-Birula.

Many of these states are unstable, but some of them appear to be stable although this has not yet been proved. (There are many definitions of stability and so one has to be very careful). As references to the assertions above, see |5,6,7|

## II. RELATIVISTIC WAVE EQUATIONS

### II.1 The nonlinear Klein-Gordon equation

There is a well-developed theory which is quite analogous to NLS, and so we just state the facts. The most basic invariant is the energy:

$$E = \int \left| \frac{1}{2} u_t^2 + \frac{1}{2}|\nabla u|^2 + \frac{1}{2} m^2 u^2 + G(u) \right| dx = \text{const} \tag{1}$$

where $G(u) = \int_0^u F(v)dv$. We consider the Cauchy problem: NLKG together with the initial (or Cauchy) conditions

$$u(x,0) = f(x) \quad , \quad u_t(x,0) = g(x)$$

The functions $f(x)$ and $g(x)$ are given functions which are arbitrary with in a certain class.

Theorem 2.1 There exists a unique solution in some time interval $|t| < t_1$, where $t_1$ may depend on the size (of some norm) of f and g.

Theorem 2.2 If $F(u) = O(|u|^2)$ as $u \to 0$, the solution can be extended to all times $-\infty < t < \infty$ provided the initial data is sufficiently small (in some norm).

This is proved by observing that, for small solutions, the term $\int G(u)dx$ in the energy E can be bounded by the term $\int u^2 dx$. Hence each term in (1) is bounded for all time.

Theorem 2.3 For arbitrary initial data of finite energy, there exists at least one weak solution for all time provided $uF(u) \geq -c(1+u^2)$ and $G(u) \geq -c$ for some $c > 0$ and all u. A major open problem is the uniqueness under these conditions.

Theorem 2.4 The solution in Theorem 2.3 is unique if $|F(u)|$ grow strictly slower than $|u|^5$ as $|u| \to \infty$. In this case the solution is as smooth as F, f and g allows.

Theorem 2.5 If $uF(u) \leq (2+\varepsilon)G(u)$ for all u and some $\varepsilon < 0$. If there exist initial data with energy $E < 0$ and $\int uu_t dx > 0$ at $t = 0$, then the solution (of Theorem 2.1) with those data blows up in a finite time.

Example 1. $F(u) = \sin u$. By Theorem 2.4 there exists a unique global $C^\infty$ solutions for arbitrary $C^\infty$ initial data.

Example 2. $F(u) = -|u|^{p-1}u$ with $p > 1$. By Theorem 2.5, many solutions blow up in a finite time.

Example 3. $F(u) = +|u|^{p-1}u$. If $1 < p < 5$, there is a unique global solution for arbitrary initial data (by Theorem 2.4). If $p \geq 5$, each pair of initial data gives rise to a global weak solution (by Theorem 2.3) but it is unknown whether this solution is unique or not. Numerical computations indicate that it is probably unique. They also show that, as p increases with the same initial data, the amplitude of the solution decreases and the number of oscillations increases. Thus the effect of a high power is to throw more of the energy into kinetic energy.

Some references for the above discussion are |8, 9, 10|.

## II.2 The Maxwell-Dirac Equations of QED

They are

$$(\mathrm{MD}) \quad \begin{cases} (i\not{\partial} - M)\psi = g \not{A} \psi \\ \dfrac{\partial}{\partial x^\nu} F^{\mu\nu} = g \bar{\psi} \gamma^\mu \psi \\ F^{\mu\nu} = \dfrac{\partial A^\mu}{\partial x_\nu} - \dfrac{\partial A^\nu}{\partial x_\mu} \end{cases}$$

The charge $\int \psi^+ \psi \, dx$ is conserved. The energy is conserved. In the Coulomb gauge

$$\sum_{k=1}^{3} \frac{\partial A^k}{\partial x_k} = 0 ,$$

it takes the form

$$e(A) = \int \left[ \frac{1}{2} \sum_{k=1}^{3} \{ ( \frac{\partial A^k}{\partial t} )^2 + |\nabla A^k|^2 \} + \frac{1}{2} |\nabla A^\circ|^2 \right] dx$$

$$e(A) + \int M\bar{\psi}\psi - \text{Im} \sum_{k=1}^{3} \left[ \partial_k \psi^+ \gamma^k \gamma^o \psi \right] dx = \text{constant}.$$

[Warning: In some other gauges, $e(A)$ will have some negative terms]. The following estimate can be deduced |11|:

$$e(A) \leq \text{const} \ (1 + \int |\nabla\psi|^2 dx)^{1/2} \qquad (2)$$

## II.3 The Yang-Mills Equations

We take the metric -+++ in space-time and the gauge group $G = SU(2)$. We write $\partial_\mu = \partial/\partial x_\mu$ ($\mu = 0,1,2,3$) for the ordinary calculus derivatives and

$$D_k = \partial_k + g \, A_k x \qquad (k = 1,2,3)$$

$$D_o = \partial_o - g \, A_o x \qquad (k = 0)$$

for the covariant derivatives. The gauge potentials $A_\mu$ are functions on space-time with values in the Lie algebra of G. Of can be regarded as ordinary 3-space and the $A_\mu$ as a vector field on space time ($\mu = 0,1,2,3$). The field strengths may be defined by the operator equations

$$g \, F_{\mu\nu} \, x = D_\mu D_\nu - D_\nu D_\mu \qquad (\mu,\nu = 1,2,3)$$

$$g \, F_{\mu o} \, x = D_o D_\mu - D_\mu D_o$$

Let $E_k = F_{ko}$ ($k = 1,2,3$), $H_1 = F_{32}$, $H_2 = F_{13}$, $H_3 = F_{21}$ be the electric and magnetic field strengths. Let E be the $3 \times 3$ matrix whose columns are $E_1, E_2$ and $E_3$. Let H be the $3 \times 3$ matrix with the columns $H_1$, $H_2$ and $H_3$. The Yang-Mills equations are the equations of motion for the Lagrangian density

$$\mathcal{L} = \frac{1}{2} |E|^2 - \frac{1}{2} |H|^2$$

The equations are (YM):

$$D_o E_1 = D_2 H_3 - D_3 H_2$$

$$D_o E_2 = D_3 H_1 - D_1 H_3$$

$$D_oE_3 = D_1H_2 - D_2H_1$$

$$D_1E_1 + D_2E_3 + D_3E_3 = 0$$

Their resemblance to Maxwell's equations is more than coincidental. In fact, they are simply Maxwell's equations suitably modified to be invariant under local gauge transformations. There are also four "constraint" equations which follow from the definitions of $D_\mu$, $E_k$ and $H_k$. They are

$$D_oH_1 = D_3E_2 - D_2E_3$$

$$D_oH_2 = D_1D_3 - D_3E_1$$

$$D_oH_3 = D_2E_1 - D_1E_2$$

$$D_1H_1 + D_2H_2 + D_3H_3 = 0$$

From these eight equations it follows that the energy is

$$\frac{1}{2}\int (|E|^2 + |H|^2)d^3x = \text{constant.}$$

See chapter III.

We shall not discuss at all the "Euclidean Yang-Mills equations" with the metric ++++, for which there has been so much recent progress. Unfortunately, on the classical level the Euclidean and Minkowski versions appear to be unrelated.

## II.4 General discussion of the existence question

The relativistic equations just discussed can all be put into the form

$$Lu = Nu \quad , \quad u(0) = \psi(x)$$

where u(0) denotes the initial data at time t=0, u is a vector function, L is a linear hyperbolic partial differential operator, and N is a nonlinear operator in u but not in the derivative of u.

One theorem is relatively easy: there is a unique solution in some time interval $|t| < t_1$ where $t_1$ depends on $\|\phi\|$. (See Theorem 2.1). There is a great deal of flexibility in choosing the norm $\|\ \|$. For MD it has been carried out in |12| and for YM in |13|. For the Einstein equations see |14|.

The global existence problem is: do arbitrary initial data $\phi$ launch solutions which exist for all time?. A possibly easier problem is to restrict the size of $\phi$. Obviously this problem must have an affirmative solution (at least for a restricted class of $\phi$) before any scattering problems can be considered. There are two methods for attaching the problem.

Method 1. Solve the equation from time 0 to time $t_1$. (For this step we need a strong enough norm $||\phi||$). Use the solution at time $t_1$ as Cauchy data and solve up to time $t_2$. Then solve up to time $t_3$ and so on. The method succeeds if the sequence of times $t_1, t_2, \ldots$ tends to infinity. Now $t_1$ depends on the size of $||\phi||$; $t_2 - t_1$ depends on the size of $||u(t_1)||$; and so on. Therefore the problem would be solved if we knew that

$$||u(t)|| \leq \text{constant}$$

for whatever times t the solution may exist. (Actually we only need to know that $||u(t)||$ is bounded for bounded t.) Mathematicians call this an a priori estimate.

Method 2. Approximate the equations. Find global solutions $u_\varepsilon$ of the approximate problems. Thus $L_\varepsilon u_\varepsilon = N_\varepsilon u_\varepsilon, u_\varepsilon(0) = \phi_\varepsilon$. Then pass to the limit and try to prove that $u_\varepsilon$ converge to a solution u of the original problem. The key step is again an a priori bound, say $|||u_\varepsilon||| \leq$ constant independent of $\varepsilon$. Then we can take weak limits $u_\varepsilon \to u$. It is usually easy to show that $L_\varepsilon u_\varepsilon \to v$, that $N_\varepsilon u_\varepsilon \to \omega$, and that $v = Lu$. Thus $Lu = w$. The difficulty in this method usually is to prove that $w = Nu$. (Easy examples show that it can happen that $w \neq Nu$). For this we again need to use a strong enough norm $|||\ \ |||$. As examples of approximation procedures we mention: (a) truncating the nonlinear terms, (b) adding viscous effects (see Prof. Dafermos's lectures), or (c) approximating by finite-dimensional problems as in numerical analysis.

They key point in both methods is the a priori estimate, which depends very precisely on the structure of the equations. The signs of the nonlinear terms are crucial, as we already saw in the cases of NLS and NLKG.

For the MD system, we know that $\int |\psi|^2 dx$ is bounded for all time. This estimate is not strong enough. The estimate (2) shows that we get a good bound on A if we could get bounds on $\int |\nabla\psi|^2 dx$. No one has found a way to accomplish this. But at least in the one-space-dimensional analogue of MD it has been carried out (see |15|).

For YM the existence question is completely open (relativistic case).

Very recently some significant progress has been made on the global existence problem for small data $\phi$ by S. Klainerman |16|. He considers essentially arbitrary (!) quadratic nonlinear perturbations of the ordinary wave equation $u_{tt} = \Delta u$, $x \in \mathbb{R}^n$. If $n \geq 5$ he proves a result like Theorem 2.2 above. The dimension enters because there is more spreading in higher dimensions. He uses the very sophisticated Nash-Moser approximation technique.

See Prof. Dafermos' lectures for further discussion of the existence question and an existence proof for some highly nonlinear systems.

## III. CONSERVATION LAWS

### III.1 The Euclidean equation

$$\Delta u = F(u(x)) \;, \quad x \in \mathbb{R}^N \tag{1}$$

We assume F is a real function such that $F(0) = 0$ and $u(x)$ is a smooth real function going to zero as $|x| \to \infty$.

Equation (1) can be written variationally as $\delta E[u] = 0$, where

$$E[u] = \int \{\tfrac{1}{2} |\nabla u|^2 + G(u)\} dx$$

is the energy and $G(u) = \int_0^u F(v)dv$. This can be expressed formally as follows. Let $T_\varepsilon$ be a smooth family of transformations such that $T_0 = I$. Let $M = dT_\varepsilon/d\varepsilon$ at $\varepsilon = 0$. For any function $u = u(x)$,

$$\frac{d}{d\varepsilon}\Big|_{\varepsilon=0} E\left[T_\varepsilon u\right] = (E'(u), Mu) = (-\nabla u + F(u), Mu).$$

Here M stands for "multiplier". If u is a solution of (1), this expression vanishes. This illustrates the general principle of Noether (1918) that if a one-parameter family of transformations leaves a variational problem invariant, the solution satisfies a conservation law. In our case it means that the product $(-\nabla u + F(u))(Mu)$ is a divergence.

It is well-known that the Laplace operator is invariant under the conformal group $\mathcal{C}$, the group of transformations on $\mathbb{R}^N$ which preserve angles. If $N \geq 3$, this group consists of four types of transformations: translations, rotations, dilation and inversions. The total dimension of $\mathcal{C}$ is therefore $N(N-1)/2 + 2N + 1$ (= 15 if N=4). On the other hand, equation (1) is invariant only under the Galilean group but not under the whole conformal group, with the exception of one particular F. We propose to exploit this fact, looking separately at the various generators of $\mathcal{C}$.

The <u>traslation</u> $T_\varepsilon : u(x) \to u(x+\varepsilon a)$, where a is a constant vector, has $M = a\cdot\nabla$ as its infinitesimal generator. Writing $(-\nabla u + F(u))(Mu)$ as a divergence, we get the conservation law

$$0 = \nabla\cdot\{-(a\cdot\nabla u)\nabla u + a(|\nabla u|^2/2 + G(u))\}$$

We get N independent laws by choosing a as the unit vector in the coordinate direction $x_k$:

$$0 = \{-u_k^2 + \frac{1}{2}|\nabla u|^2 + G(u)\}_k + \sum_{j\neq k}\{-u_j u_k\}_j \tag{2}$$

where subscripts denote partial derivatives.

The <u>rotations</u> give the $N(N-1)/2$ multipliers $x_k u_j - x_j u_k$ for $j \neq k$ and the conservation laws

$$0 = \nabla\cdot\{(-x_k u_j + x_j u_k)\nabla u\} + \{x_k(|\nabla u|^2/2 + G(u))\}_j - \{x_j(|\nabla u|^2/2 + G(u))\}_k. \tag{3}$$

The <u>dilation</u> $u \to u_\lambda$ leaves the Dirichlet integral invariant, where $u_\lambda(x) = \lambda^m u(\lambda x)$. To find the correct value of m, we calculate $\nabla u_\lambda(x) = \lambda^{m+1}(\nabla u)(\lambda u)$ and

$$E[u_\lambda] = \int\{\frac{1}{2}\lambda^{2m+2}|(\nabla u)(\lambda x)|^2 + G(\lambda^m u(\lambda x)\}dx = \int\{\frac{1}{2}\lambda^{2m+2-N}|\nabla u(y)|^2 + \lambda^{-N}G(\lambda^m u(y))\}dy$$

where $y = \lambda x$, $dy = \lambda^N dx$. The first term is invariant if $2m+2-N = 0$ or $m = (N-2)/2$. For this choice of m,

$$0 = \frac{d}{d\lambda} E[u_\lambda]\Big|_{\lambda=1} = \int\{-NG(u) + muF(u)\}dy. \tag{4}$$

The multiplier is

$$Mu = \frac{d}{d\lambda}\lambda^m u(\lambda x)\Big|_{\lambda=1} = x\cdot\nabla u + mu$$

The conservation law is

$$0 = \frac{N-2}{2}uF(u) - NG(u) + \nabla\cdot\{(x\cdot\nabla u)\nabla u + \frac{1}{2}x|\nabla u|^2 + \frac{N-2}{2}u\nabla u + xG(u)\} \tag{5}$$

Equation (7) provides some non-trivial information about possible solutions of (1). We have

$$\int|\nabla u|^2 dx = -\int uF(u)dx = \frac{-2N}{N-2}\int G(u)dx$$

(if N $\neq$ 2). Therefore

$$E|u| = \frac{1}{N}\int |\nabla u|^2 dx \geq 0$$

The following theorem follows easily.

Theorem 3.1 If u is a solution of (1), smooth and zero at infinity, then the energy is positive (except if $u \equiv 0$). There can be no solution of (1) if any one of the following four functions is positive (for $s \neq 0$):

$$sF(s),\ G(s),\ H(s),\ -H(s)$$

where we assume $N \neq 1$, $H(s) = (N-2)sF(s) - 2NG(s)$.

The one-dimensional case (N = 1) is truly exceptional since it permits solutions even if $G \geq 0$. The theorem is due in part to Derrick |7| and in part to Strauss |5|.

We have seen above that the nonlinear equation is not invariant under the transformation $u \to u_\lambda$. However, it is invariant in the special case

$$-NG(u) + \frac{N-2}{2} uF(u) = 0,\ G' = F.$$

That is, $G(u) = \text{const } u^{2N/(N-2)}$. In this case, our variational problem is equivalent to finding the best Sobolev constant $\|\phi\|_{2N/(N-2)} \leq \text{const } \|\nabla\phi\|_2$. See Strauss |5|.

The inversion $V: x \to x/x{\cdot}x$ is the forth kind of conformal transformation. It leaves the unit sphere $|x|^2 = 1$ invariant and $V^2 = I$. If we let $v(x) = |x|^{2-N}u(x|x|^{-2})$, a calculation shows that $\int |\nabla v(x)|^2 dx = \int |\nabla u(y)|^2 dy$. An N-parameter family of inversions is given by $y = V_a(x)$ where

$$y/|y|^2 = x/|x|^2 + a \quad (a \in \mathbb{R}^N)$$

That is, $V(y) = T_a V(x)$ where $T_a$ is translation. So we may write $V_a = VT_aV$ or

$$y = V_a(x) = \frac{x + a|x|^2}{1 + 2a{\cdot}x + |a|^2|x|^2}$$

These inversions given us N rather complicated conservation laws. The multipliers are essentially

$$\frac{\partial}{\partial\varepsilon} u(V_{\varepsilon a}(x))\Big|_{\varepsilon=0} = |x|^2 a.\nabla u - 2(a.x)(x.\nabla u)$$

Precisely, they are

$$Mu = \left[|x|^2 a - 2x(x-a)\right]\cdot\nabla u - (N-2)(x\cdot a)u \tag{6}$$

We leave as an exercise the writing of the inversional conservation laws.

III.2 NLKG.

We take n space dimensions. We can transfer each of the Euclidean identities by making the following changes:

$$N = n+1,$$
$$x \to (x_1, x_2, \dots, x_n, it),\ x_N = x_{n+1} = it$$
$$F(u) \to m^2 u + F(u)$$

Thus there are $(N^2 + 3N + 2)/2$ identities which immediately follow from their Euclidean counterparts. Here they are, after integration over space coordinates only.

From the multiplier $u_t = \partial_t u$, we get the energy

$$\int e(u)dx = \int (\tfrac{1}{2} u_t^2 + \tfrac{1}{2} |\nabla u|^2 + \tfrac{1}{2} m^2 u^2 + G(u)dx = \text{constant}.$$

From the multiplier $u_k = \partial_k u$, we get the momenta

$$\int u_t u_k \, dx = \text{constant}.$$

We get the angular momenta from the multipliers $x_k u_t + t u_k$ and $x_k u_j - x_j u_k$:

$$\int (x_k e(u) + t u_k u_t) dx = \text{const}.$$

and

$$\int (x_k u_j - x_j u_k) u_t dx = \text{const}.$$

The next two identities are due to Morawetz |12|: From the multiplier $Mu = tu_t + ru_r + \frac{n-1}{2} u$, where $r = |x|$ is the spatial radius, we get the dilational identity

$$0 = \frac{d}{dt} \int (te(u) + r u_r u_t + \tfrac{n-1}{2} u u_t) dx + \tfrac{1}{2} \int H(u) dx$$

where

$$H(u) = (n-1)uF(u) - 2(n+1)G(u) - 2m^2u^2$$

Finally we get the inversional or conformal identities. From the multiplier ($k = N = n+1$ in (16))

$$Mu = (t^2 + r^2)u_t + 2rtu_r + (n-1)tu,$$

we get the identity

$$0 = \frac{d}{dt}\int\Big[(t^2+r^2)e(u) + 2rtu_r u_t + (n-1)tuu_t - \frac{n-1}{2}u^2\Big]dx + t\int H(u)dx \tag{7}$$

From the multiplier (see (6))

$$Mu = tx_k u_t + \frac{1}{2}(t^2 + 2x_k^2 - r^2)u_k + x_k \sum_{j\neq k} x_j u_j + \frac{n-1}{2}x_k u,$$

we get the identity

$$0 = \frac{d}{dt}\int\Big[tx_k e(u) + \frac{1}{2}(t^2 + 2x_k^2 - r^2)u_k u_t + x_k \sum_{j\neq k} x_j u_j u_t + \frac{n-1}{2}x_k u_t u\Big]dx + \frac{1}{2}\int x_k H(u)dx$$

Another identity due to Morawetz |19| is obtained using the spatial radial derivative as the multiplier. Thus

$$Mu = \frac{\partial u}{\partial r} + \frac{n-1}{2r}u, \quad r = |x| .$$

A direct calculation shows

$$0 = \frac{d}{dt}\int u_t(u_r + \frac{n-1}{2r}u)dx + \int(|\nabla u|^2 - u_r^2)\frac{dx}{r} + \frac{(n-1)(n-3)}{4}\int u^2\frac{dx}{r^3} +$$

$$\frac{n-1}{2}\int(uF(u) - 2G(u))\frac{dx}{r} \tag{8}$$

for $n \geq 3$, with the extra term $2\pi u^2(0,t)$ in case $n = 3$.

## III.3 The Yang-Mills equations

They too are conformally invariant (as are Maxwell's equations), so again there are 15 conservation laws. One fo them is the energy already written in chapter II. The momenta are $\int p^k dx$ = constant, where

$$p^1 = H^2 \cdot E^3 - E^2 \cdot H^3$$

and $p^2$, $p^3$ defined similarly. Of the other conservation laws the most useful one is the inversional law analogous to (7). For YM it takes the form

$$\iint\left[(t^2 + r^2)\,\frac{1}{2}\,(|E|^2 + |H|^2) + 2t\sum_k x_k p^k\right]dx = \text{constant} \tag{9}$$

If we complete the square in this integral, we see that it is at least

$$(t - r)^2(|E|^2 + |H|^2)/2$$

which is positive! In particular, if we restrict the integral to any cone of smaller aperture than a light cone, we obtain the following result |20|.

Theorem 3.2 Let $R > 0$ and $0 < \varepsilon \leq 1$. As $t \to \infty$ we have

$$\int_{|x|<R+(1-\varepsilon)t} (|E|^2 + |H|^2)\,dx = O(t^{-2})$$

This theorem means that all the energy moves along the light cone; that is, at unit speed. It follows that YM cannot possess a solution of the form $E(x - ct)$, $H(x - bt)$ with $|b| < 1$, $|c| < 1$ which vanishes at infinity. Thus we can say that YM possesses no generalized bound state. It can also be shown that 12 of the 18 components of a finite energy solution of YM are square integrable on any light cone. For the details, see |20|. The same result is true for a Yang-Mills field coupled to a Klein-Gordon field as in a Higgs model.

The Maxwell-Dirac equations in the mass-zero case are also conformally invariant. The resulting conservation laws have been calculated in |11|.

## III.4 NLS

We assume $F(u) = g'(|u|)u/|u|$ where $g(0) = g'(0) = 0$ and we let $G(u) = g(|u|)$. It has the Lagrangian

$$\iint\left[-\mathrm{Im}(u_t\bar{u}) + \frac{1}{2}\,|\nabla u|^2 + G(u)\right]dxdt$$

which is invariant under the transformation $u(x,t) \to \lambda^{n/2}u(\lambda x,\lambda^2 t)$ if $G = 0$. This leads to the multiplier $r\bar{u}_r + 2t\bar{u}_t + (n/2)\bar{u}$ and the dilational identity:

$$\frac{d}{dt}\int\left[\frac{1}{2}\,\mathrm{Im}(ru_r\bar{u}) + t|\nabla u|^2 + 2tG(u)\right]dx + \frac{1}{2}\int\left[nF(u)\bar{u} - 2(n+2)G(u)\right]dx = 0 \tag{10}$$

Ginibre and Velo |21| discovered the following analog of the inversional identity , which they call the pseudo-conformal conservation law. For the free evolution operator,

$$xU_o(-t)f = U_o(-t)(x + 2it\nabla)f.$$

The identity is

$$\frac{d}{dt}\int\left[\frac{1}{2}\,|xu - 2it\nabla u|^2 + 4t^2G(u)\right]dx + 2t\int\left[n\bar{u}F(u) - 2(n+2)G(u)\right]dx = 0 \tag{11}$$

Let $F(u) = \lambda|u|^{p-1}u$. If $p = 1 + 4/n$, both (2) and (3) are exact conservation laws. If $\lambda > 0$, (11) implies that $\int G(u)dx = O(t^{-2})$.

## IV. NONLINEAR SCATTERING THEORY

IV.1 In the absence of bound states, one normally expects the asymptotic behavior as $t \to \pm\infty$ to be free. Thus if $u(t) = u(x,t)$ is the interacting field, we look for free fields $u_+(t)$ and $u_-(t)$ such that

$$||u(t) - u_\pm(t)|| \to 0 \quad \text{as } t \to \pm\infty \tag{1}$$

The scattering operator S is defined by

$$S:\ u_-(0) \to u_+(0)$$

For illustration let us take the NLS equation

$$i\,\frac{\partial u}{\partial t} - \Delta u + F(u) = 0 \qquad (x \in \mathbb{R}^n)$$

where $\arg F(u) = \arg u$. Let $H_0 = +\Delta$ be the free Hamiltonian. Then we define (formally for the time being)

$$u_-(t) = u(t) + \int_{-\infty}^{t} e^{i(t-s)H_0}\,F(u(s))ds \tag{2}$$

Formally $u_-$ satisfies the asymptotic property (1). By differentiation of (2) we have

$$\frac{\partial u_-}{\partial t} = \frac{\partial u}{\partial t} + iH_0\int_{-\infty}^{t}\ldots + F(u(t))$$

$$= \frac{\partial u}{\partial t} + iH_0(u_- - u) + F(u) = iH_0u_-$$

which is the free Schrödinger equation. We consider it to action the Hilbert space $\mathcal{H} = L^2(\mathbb{R}^n)$.

As a second illustration, take NLKG. It can be written in Hamiltonian form as follows. Let

$$\vec{u} = \begin{pmatrix} u \\ u_t \end{pmatrix} \quad , \quad iH_o = \begin{pmatrix} 0 & 1 \\ \Delta - m^2 & 0 \end{pmatrix} \quad , \quad \vec{F} = \begin{pmatrix} 0 & 0 \\ F & 0 \end{pmatrix}$$

Then NLKG is equivalent to the "vector" equation

$$\frac{\partial \vec{u}}{\partial t} = iH_o\vec{u} - \vec{F}(\vec{u})$$

The free Klein-Gordon equation is equivalent to $\partial\vec{v}/\partial t = iH_o\vec{v}$. It acts on the energy Hilbert space $\mathcal{H}$. The norm is

$$||\vec{u}||^2 = \int (u_t^2 + |\nabla u|^2 + m^2 \; |u|^2)dx$$

Thus the first component u of $\vec{u}$ belongs to $L^2$ and its spatial derivatives as well; the second component $u_t$ of $\vec{u}$ also belongs to $L^2$. The solution of the free equation $\exp(itH_o)$ is a unitary operator on $\mathcal{H}$. The generator $iH_o$ is a self-adjoint operator (but not taking all of $\mathcal{H}$ into $\mathcal{H}$). The interaction $\vec{F}$ is a nonlinear operator which may take all of $\mathcal{H}$ into $\mathcal{H}$ under special circunstances.

With this notation equation (2) is the definition of $\vec{u}_-$ if arrows (→) are put on $\vec{F}$ and $\vec{u}$. This is a 2 component integral equation. If we write its first component, it takes the form

$$u_-(t) = u(t) - \int_{-\infty}^{t} D_{ret}(t-s) \quad F(u(s))ds$$

Here $D_{ret}$ is the retarded Green's function for the KG equation. This is known as the Yang-Feldman equation.

IV.2 Mathematical results.

The discussion which follows focuses on NLKG. The integral in (2) converges absolutely in the energy norm (norm of $\mathcal{H}$) provided

$$\int_{-\infty}^{0} (\int |F(u(x,t))|^2 dx)^{1/2} dt < \infty$$

This leads naturally to the question of the decay as $|t| \to \infty$ of the solutions of NLKG. Now it is well known that free solutions (with sufficiently nice initial data) decay to zero like $|t|^{-3/2}$. Is the same true for the solutions of NLKG? Under certain circunstances, yes.

Theorem 4.1 (Low-Energy Scattering). Take n=3. Let $|F(u)| = O(|u|^3)$ as $u \to 0$. Let $u_-(x,t)$ be any sufficiently small free solution (or else assume a sufficiently small coupling constant) of finite energy satisfying

$$\max_x |u_-(x,t)| = O(|t|^{-3/2}) \tag{3}$$

Then there exists a unique solution $u(x,t)$ of NLKG and a unique free solution $u_+(x,t)$, both of which are also small, which satisfy (1). The 3 solutions $u_-, u, u_+$ are related by the Yang-Feldman equations.

Let $\Sigma$ be the space of initial data $\vec{u}_-(0)$ of solutions $u_-(x,t)$ satisfying $\vec{u}_-(0) \in \mathcal{H}$ and (3). Then the scattering operator S, which takes $\vec{u}_-(0)$ into $\vec{u}_+(0)$, maps a neighborhood of the zero solutions in $\Sigma$ into $\Sigma$.

Examples: This theorem is applicable to $F(u) = \pm u^p$ with $p \geq 3$. Also $F(u) = \sin u - u$. The conclusion has been proved false if $1<p<1+2/3$.

The proof of Theorem 4.1 is not difficult (see |8|). It is based on an iteration procedure (Picard method or Born approximation) familiar to both physicists and mathematicians.

The <u>inverse scattering</u> problem is to determine the interaction F from the scattering operator S. Under the conditions of Theorem 4.1, this can be done. In fact, we define

$$B(u,v) = \int (uv_t - u_t v)dx.$$

We calculate

$$\frac{d}{dt} B(u,v) = \int (uF(v) - F(u)v)dx$$

Hence

$$B(u_+,v_+) - B(u_-,v_-) = \iint (uF(v) - F(u)v)dx\, dt \tag{4}$$

The left side of (4) is determined by S and a pair of arbitrary inputs $u_-, v_-$. Let $u_-$ and $v_-$ are small, say $u_- = \varepsilon\phi$ and $v_- = \varepsilon\psi$. Then the right side of (4) can be expanded in powers of $\varepsilon$. If F is a power series (analytic function), then each coefficient can be successively determined. Even if F depends on x, it can be determined. For instance if $F(x,u) = V(x)u^3$, then the "potential" $V(x)$ can be recovered from S(see |18,22,23|).

Theorem 4.1 is limited to the consideration of small solutions. But large ones are physically more interesting. We already saw in chapter II some of the difficulties which can occur when p is large.

<u>Theorem 4.2</u> Let $F(u) = u^3$, $n = 3$. Then there is another space $\mathcal{F}$ such that

$$\Sigma \subset \mathcal{F} \subset \mathcal{H}$$

and $$S: \mathcal{G} \to \mathcal{G}$$

Every solution u(x,t) of NLKG with initial data in $\Sigma$ also decays as in (3).

The proof of this theorem is long and mathematically tricky. But its idea is simple. We go back to equation (8) of chapter III whence we observe that

$$\iint u^4 \frac{dx}{r} \, dt < \infty \tag{5}$$

This estimate comes from integrating (8) over all time and bounding the first term by the energy. We may assume u(x,t) vanishes outside a light cone $|x| > t+k$ and thus

$$\int^{\infty} f(t)dt/t < \infty \quad \text{where } f(t) = \int u^4 dx$$

This is an extremely weak statement of decay. Because $t^{-1}$ is not integrable, f(t) could not be a constant and in fact $\int_I f(t)dt$ is arbitrarily small on arbitrarily long time intervals I. The proof fo decay continues like a jacking-up process. The succeeding steps are that: u(x,t) is arbitrarily small on arbitrarily long time intervals; $u(x,t) \to 0$ uniformly as $t \to \infty$; $\sup_x |u(x,t)|^2$ is integrable, and finally $|u(x,t)| = O(t^{-3/2})$.

The most interesting step is the uniform convergence to zero. Let $\varepsilon$ be a positive number. Let $T = T(\varepsilon)$ be sufficiently large. By the preceding step, $|u(x,t)| < \varepsilon$ on some time interval $|t^* - T, t^*|$. Let

$$t^{**} = \sup\{ s \mid |u| < \varepsilon \quad \text{in} \quad |t^* - T, s| \}$$

If $t^{**} = \infty$, there is nothing to prove. Suppose $t^{**} < \infty$. Take a time t slightly later than $t^{**}$; namely $t^{**} \le t \le t^{**} + \delta$. Break up the right side of (2) into four parts. Since $t \ge t^{**} \ge T$ is large enough, $|u_0| < \varepsilon/4$. The integral over $|t^{**}, t|$, the tip of the cone, is less than $\varepsilon/4$ if $\delta$ is chosen small enough. In the interval $|t-T, t^{**}|$, we have $|u(x,t)| < \varepsilon$. Since u appears in (2) to the third power and $\varepsilon$ is small, we can arrange the integral over $|t-T, t^{**}|$ to be less than $\varepsilon/4$, no matter how large T is. The fourth part is over the large base of the cone $|0, t-T|$, where we do not know that u is small. However $t-s > T$ in that interval and so $R(x-y, t-s)$ is small in some sense. The kernel is actually constant on the hyperboloids $\mu$ = constant, but they bunch together very closely and contribute little to the integral. Altogether, we obtain $|u(x,t)| < 4(\varepsilon/4) = \varepsilon$, which contradicts the definition of $t^{**}$. This proves the uniform decay. For the details of this proof, see |23|.

One can prove the following properties |23| enjoyed by S.

(a) S maps $\mathcal{F}$ one-one onto $\mathcal{F}$

(b) S is a diffeomorphism on $\mathcal{F}$

(c) S is Lorentz-invariant.

(d) S commutes with the free group exp it $H_0$.

(e) S is odd.

(f) $\|Sf\| = \|f\|$ (energy norm)

(g) S is not a linear operator

Theorem 4.1 is rather easily generalized. It has the same character as do Theorem 2.2 and Klainerman's new result mentioned at the end of chapter II. Theorem 4.2 is more delicate. It generalizes to (i) the mass-zero case still with n=3 and to (ii) NLS in any dimension. Actually these two cases are considerably easier than NLKG because of the conservation laws: equation (7) of chapter III for (i) and equation (11) of chapter III for (ii). In either case we have

$$\int G(u)dx = O(t^{-2})$$

which is a far stronger estimate than (5). In the case of NLS the analogue of Theorem 4.2 is valid if $F(u) = |u|^{p-1}u$ if

$$1 + 4/n \leq p < 1 + 4/(n-2)$$

These two special values of p are familiar form our easlier discussions. This result was proved in the elegant papers of Ginibre and Velo |21| using the conservation law which they discovered. For n = 3 it was also proved by Lin and Strauss [J. Funct. Anal, 1978] who generalized the method used for NLKG.

## IV.3 Scattering of arbitrary finite-energy solutions

In standard linear scattering theory the total Hamiltonian generates a unitary group of operators and so S too is a unitary operator. If S is defined on a dense set in $\mathcal{H}$, it automatically extends to all of $\mathcal{H}$. In our nonlinear problems S is defined on a dense subset $\mathcal{F}$ of the Hilbert space $\mathcal{H}$. Is it possible that S map all of $\mathcal{H}$ into $\mathcal{H}$?. A partial result in this direction is the following.

Theorem 4.3 Consider NLKG with m > 0 in any dimension n. Assume $F(u) = |u|^{p-1}u$ where

$$1 + 4/n \leq p \leq 1 + 4/(n-1)$$

Let $u_-(x,t)$ be <u>any</u> free solution of finite energy. (a) Then there exists a solution $u(x,t)$ of NLKG such that $\|u(t) - u_-(t)\| \to 0$ as $t \to -\infty$ (b) If $\|u_-(0)\|$ is sufficiently small, then there exists a free solution $u_+(x,t)$ such that $\|u(t) - u_+(t)\| \to 0$ as $t \to +\infty$. (c) If $\iint |u|^{p+1} dxdt$ is finite, then $n_+(x,t)$ exists.

We conjecture that the hypothesis in (c) is always true. If that were so, the scattering operator S would be a mapping of all <u>of</u> $\mathcal{H}$ into $\mathcal{H}$.

The proof of Theorem 4.3 is based on the functional-analytical techniques of Ginibre and Velo and on some new <u>a priori</u> decay estimates for the <u>free</u> KG equation. We now state these estimates.

Consider a free solution

$$v_{tt} - \Delta v + m^2 v = 0 \quad , m > 0, \quad x \in \mathbb{R}^n$$

$$v(x,0) = 0 \quad , \quad v_t(x,0) = g(x)$$

Its energy is finite if $g \in L^2$. Consider the mapping $g \to v$. The first estimate |24| is

$$|v(x,t)|^q \, dxdt \leq \text{const} \left(\int g^2 dx\right)^{q/2} \tag{6}$$

if $2+4/n \leq q \leq 2+6/(n-2)$ ($q < \infty$ if $n = 1,2$). In particular, <u>all</u> finite-energy solutions decay in some sense as $|t| \to \infty$. The second estimate |25| is

$$\int |v(x,t)|^q \, dxdt \leq \frac{\text{const}}{t^2} \left(\int |g(x)|^{q'} dx\right)^{q-1}$$

where $q' = q/(q-1)$ and $2+4/n \leq q \leq 2+4(n-1)$. In order to prove them, one makes a Fourier decomposition

$$v(x,t) = c \int e^{ix\cdot k}(m^2 + k^2)^{-1/2} \sin \left| t(m^2 + k^2 \right|^{1/2} \hat{g}(k)dk$$

and then uses techniques of Fourier analysis and the interpolation theory of operators.

Theorem 4.3 uses these estimates with $q = p+1$. Part (a) is proved in |26|. If $\|u_-(0)\|$ is small, then $\iint |u_-|^{p+1} dxdt$ is also small by (6). Using an iteration procedure, we deduce that $\iint |u|^{p+1} \, dxdt$ is finite. Thus (b) is reduced to (c). Part (c) is proved by the techniques of |26| although the proof is not given there. As for our conjecture, there is considerable evidence to support it: see estimate (5).

## IV.4 Inverse Scattering

We have already mentioned the nonlinear inverse scattering problem. One of the most exciting developments in recent years in nonlinear partial differential equations is the theory of solitons |27|. The key discovery was that certain highly nonlinear problems can be reduced to linear ones, in particular to the inverse scattering problem for the linear Schrödinger equation. Its main limitation at present is that the technique is successful only in one dimension. The tool used by everyone has been the celebrated Gelfand-Levitan equation.

Very recently Deift and Trubowitz |28| have formulated a new approach to this problem which provides insight into its structure and hope of its extension to higher dimensions. They consider the Hamiltonian

$$H = - d^2/dx^2 + V(x) \quad , \quad - \infty < x < \infty$$

there are two points at infinity ($\pm\infty$) so that the scattering operator is a $2 \times 2$ matrix in this case. Let $f_+$ and $f_-$ be the solutions of $Hf = k^2 f$ with $f_\pm \sim \exp(\pm ikx)$ as $x \to \pm\infty$. One easily shows that there are functions $T, R_+$ and $R_-$ such that

$$T(k)\ f_\pm(x,k) \sim e^{\pm ikx} + R_\pm(k)\ e^{\mp ikx}$$

as $x \to \mp\infty$. Then S turns out to be the matrix

$$S(k) = \begin{pmatrix} T(k) & R_+(k) \\ R_-(k) & T(k) \end{pmatrix}$$

Let us take the case when H has no bound states. The problem is to recover the potential $V(x)$ from the reflection coefficient $R_+(k)$. Assume $V(x) \to 0$ sufficiently fast as $|x| \to \infty$. The key new idea is the formula

$$V(x) = \frac{2i}{\pi} \int_{-\infty}^{\infty} k\ R_+(k)\ f_+^2(x,k) dk \tag{7}$$

They write the differential equation in the form

$$- f_+''(x,l) + V(x)\ f_+(x,l) = l^2 f_+(x,l)$$

with $V(x)$ replaced by (7). This is an infinite system of ordinary differential equations which are coupled through all the frequencies k in the integral. They solve it by simple iteration for $f_+(x,k)$ and recover $V(x)$ by formula (7).

## REFERENCES

|1| R.Y. Chias, E. Garmire and C.H. Townes, Phys. Rev. Lett. 13 (1964), 479-482.

|2| P.L. Kelley, Phys. Rev. Lett. 15 (1965), 1005-1012;
V.I. Talanov, JETP Lett. 2 (1965), 138-141.

|3| J.B. Baillon, T. Cazenave and M. Figueira, C.R. Acad. Sci. 284 (1977), 869-872.

|4| W.A. Strauss, Anais Acad. Brasil. Ciencias 42 (1970), 645-651.

|5| W.A. Strauss, Comm. Math. Phys. 55 (1977), 149-162.

|6| W.A. Strauss and L. Vázquez, to appear.

|7| H. Berestycki and P.L. Lions, to appear.

|8| W.A. Strauss, in Invariant Wave Equations, Erice 1977, Springer Lect.Notes in Physics no.73, p.197-249.

|9| W.A. Strauss and L. Vázquez, J. Comp. Phys. 28, (1978), 271-278.

|10| M. Reed, Abstract Nonlinear Wave Equations, Springer Lect. Notes in Math. No. 507, 1976.

|11| R.T. Glassey and W.A. Strauss, Conservation laws for the classical Maxwell-Dirac and Klein-Gordon-Dirac equation, J. Math. Phys., to appear.

|12| L. Gross, Comm. Pure Appl. Math. 19 (1966), 1-15.

|13| I.E. Segal, The Cauchy problem for the Yang-Mills equations, to appear.

|14| J. Hughes, T. Kato and J. Marsden, Arch. Rat. Mech. Anal. 63 (1977), 273-294.

|15| J.M. Chadam, J. Funct. Anal. 13 (1973), 173-184;
R.T. Glassey and J.M. Chadam, Proc. A.M.S. 43 (1974), 373-378.

|16| S. Klainerman, Global existence for nonlinear wave equations, Ph.D. dissertation, New York Univ. 1978.

|17| G.H. Derrick, J. Math. Phys. 5 (1964), 1252.

|18| C.S. Morawetz, Comm. Pure Appl. Math. 15 (1962), 349-362;
C.S. Morawetz, Notes on Time Decay and Scattering for some Hyperbolic Problems, Soc. Industr. Appl. Math. Philadelphia, 1975.

|19| C.S. Morawetz, Proc. Roy. Soc. A306 (1968), 291-296.

|20| R.T. Glassy and W.A. Strauss, Decay of Classical Yang-Mills fields, Comm. Math. Phys.; also, Decay of a Yang-Mills field coupled to a scalar field, to appear.

|21| J. Ginibre and G. Velo, On a class of nonlinear Schrödinger equations, to appear.

|22| W.A. Strauss, in Scattering Theory in Math. Phys. Reidel Publ. Co., Dordrecht, 1974, 53-78.

|23| C.S. Morawetz and W.A. Strauss, Comm. Pure Appl. Math. 25 (1972), 1-31 and 26 (1973), 47-54.

|24| I.E. Segal, Adv. Math. 22 (1976), 305-311;
R.S. Strichartz, Duke Math. J. 44 (1977), 705-714.

|25| B. Marshall, W. Strauss and S. Wainger, to appear.

|26| W.A. Strauss, in Nonlinear Evolution Equations, North-Holland (1978), to appear.

|27| A.C. Scott, F.Y.F. Chu, and D.W. McLaughlin, Proc. IEEE 61 (1973), 1443-1483.

|28| P. Deift and E. Trabowitz, Comm. Pure Appl. Math., to appear.

This work was supported in part by NSF Grants MCS 75-08827 and MCS 78-03567.

NON-LINEAR TRANSPORT EQUATIONS :

PROPERTIES DEDUCED THROUGH TRANSFORMATION GROUPS

---

J. GUTIERREZ*, A. MUNIER**, J.R. BURGAN,
M.R. FEIX, E. FIJALKOW***
CRPE/CNRS, UNIVERSITE D'ORLEANS (FRANCE)

ABSTRACT :

Transport equations in configuration space (linear and non-linear heat equations) and in phase space (Vlasov-Poisson systems for plasmas, beams and gravitating gases) are considered in the frame of transformation group techniques. Both self-similar and more general groups are introduced to find specially interesting solutions. Two kinds of results are obtained : time evolution of given initial situations and systematic derivation of possible scaling laws for a given mathematical model. These last results are specially interesting for extrapolating performances of Fusion Machines.

* Supported by C.I.E.S. France. Permanent address Universidad de Alcala de Henares Madrid (SPAIN).

** Compagnie Internationale de Services en Informatique Paris.

*** U.E.R. Sciences

## I. INTRODUCTION

In the last few years, non-linear problems have become of major interest in theoretical and experimental Plasma Physics as in other scientific areas. Nevertheless, after much hard work by several authors who have developped a number of analytical[1] [2] [3] and numerical[4] [5] methods to solve non-linear plasma equations, only a few quasi-linear problems have been investigated.

We intend to present here some ideas which can suggest new ways to study different phenomena, not only in Plasma Physics, but in other branches of Mathematical Physics.

Firstly we shall look at the applications of transformation group theory to the solution of non-linear partial differential equations. This technique was propounded by Sophus Lie[6] at the end of the nineteenth century and it has been applied in the last decade[7] [8] [9] [10].

We sketch quickly this method ; consider a non-linear (or linear) partial differential system of equations with n independent variables. Find a group of transformation which keeps this system formally invariant*. Form those particular combinations of the independent and dependent variables which stay invariant under the transformation, and call them the "invariants". Now, any function which depends on the "invariants" only, is obviously invariant itself. Then, look for particular solutions of the system, as particular functions of the invariants. After substitution into the PDEs obtain a new system of equations that these particular solutions must satisfy. This system is the "reduced system" and usually contains less than n new independent variables. Evidently, due to this smaller number of independent variables, the new system is often easier to solve and in some cases analytical solutions can be found, but it will ordinarily be also non-linear. On the other hand, if the problem is time dependent and if it is possible to "eliminate" (it would be more correct to say to absorb) the t-variable, the solutions of the reduced system involve the whole temporal behaviour of the corresponding initial conditions.

* (i.e. the equations remain identical except that new variables have replaced the old ones)

For the time dependent problems most of the solutions obtained by this method correspond to unphysical initial conditions. A similar difficulty is found for stationary problems in relation to boundary conditions. However, for some cases one can obtain the analytical and total time behaviour of some subsets of the physical system[11].

Our philosophy consists in studying thoroughly a problem from the analytical point of view, in order to simplify it, using the transformation group technique as far as possible, and to return to numerical methods if we are compelled to by the nature of the solution.

To sum up, we give priority to the analytical methods and often we work in a double analytical and numerical frame.

In paragraph II the implementation rules of the transformation group technique are explained together with the connection between self-similar and infinitesimal groups, with the non-linear heat equation as an example.

We are mainly interested in phase space fluids, which are defined in Section III, where we give some comments about their invariant solutions. Another important application of self-similar groups is shown in IV : It is the possibility to obtain scaling laws for experimental devices (Fusion machines in this case). We are not interested in the invariant solutions of this problem, but simply in the group invariants which will give the scaling between technological variables wherever possible (Here we will "project" the experimental knowledge of one machine to build a "family" of new improved devices).

Due to the difficulty of finding physically invariant solutions, the idea of quasi-invariance transformation groups is introduced in paragraphs V and VI. The technique consists in simplifying the problem as much as possible from the analytical point of view. The objective is the renormalization of time and forces in order to simplify numerical calculations and obtain information on asymptotic behaviour. We shall give the connection between these transformations and those of analytical mechanics.

It must be said that the existence of the above group of transformation (invariance and quasi-invariance groups) is not always possible and other groups of transformations ought to be investigated.

## II. TRANSFORMATION GROUPS AND FORMAL INVARIANCE

### a) Infinitesimal group technique

We are going to give some tedious but necessary technical details and precise the notation.

Let L = 0, a system of partial differential equations in principle non-linear. The $u_1 \dots u_m$, are the dependent variables and the $x_1 \dots x_n$ the independent ones.

An infinitesimal group of transformations G, with parameter $\varepsilon \ll 1$

|1|

$$\overline{x}_1 = x_i + \varepsilon A_i\ (x_j,\ u_j) + \theta(\varepsilon^2)$$

$$\overline{u}_i = u_i + \varepsilon B_i\ (x_j,\ u_j) + \theta(\varepsilon^2)$$

is chosen in order to keep the system formally invariant to the first order in $\varepsilon$. It makes possible the determination of the arbitrary functions $A_i$, $B_i$ if they exist in a non-trivial form.

Now, we look for those functions of $x_i$ and $u_i$ which are invariant under the general group G or under any subgroup $H \subset G$.

That is to say, the functions $J(x_i,\ u_i)$ which verify if $T_g \in G$

|2| $$T_g\ J(x_i,\ u_i) \equiv \overline{J}(\overline{x_i},\ \overline{u_i}) \equiv J(\overline{x_i},\ \overline{u_i}) = J(x_i,\ u_i)$$

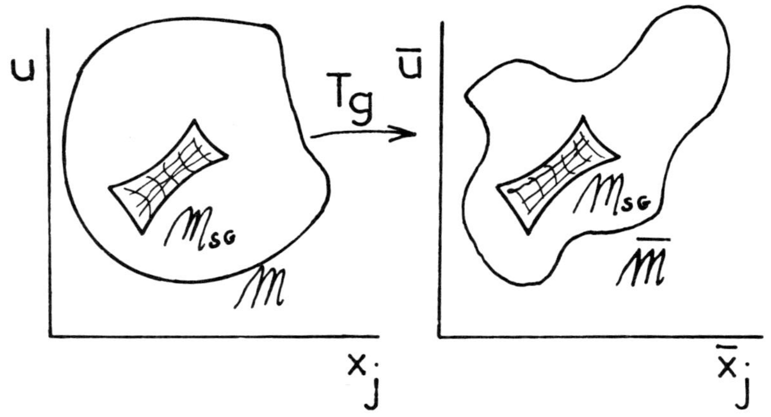


FIGURE 1

Transformed manifold $\bar{M}$ = Tg M under an element Tg of the total invariance group G. The manifold M is generated by the solutions of a system of partial differential equations.
$M_{sg}$ in the diagramme corresponds to the invariant manifold under the group of transformations generated by the invariant solutions.

Let n be the number of independent variables, there will be $m < n$ independent invariants. Evidently an arbitrary function of these invariants is itself invariant.

In order to find invariants, the corresponding group or subgroup operators must be used. They are defined in the form

$$|3| \qquad X_{tg} = \Sigma_i \left(A_i \frac{\partial}{\partial x_i} + B_i \frac{\partial}{\partial u_i}\right)$$

where $A_i$ and $B_i$ are the generators of the corresponding subgroup. In this representation an invariant function will be

$$X_{Tg}\, J(x_i, u_i) \equiv J(\overline{x_i}, \overline{u_i}) = J(x_i, u_i)$$

Obeying in practice the following partial differential equations

$$|4| \qquad \Sigma_i \left(A_i \frac{\partial J}{\partial x_i} + B_i \frac{\partial J}{\partial u_i}\right) = 0$$

The way to obtain invariant solutions consists in assuming that the invariants are the characteristics of Eq. 4. These solutions generate the whole manifold remaining invariant under the transformations Tg (Fig. 1).

If these invariant curves of the manifold are solutions of the system of partial differential equations, they must satisfy it. It is in this way that the reduced system is found since the knowledge of a set of invariants allows a decrease of the rank of the system.

b) Example : The non-linear heat diffusion equation

Let us start with the one dimensional heat equation with two independent variables x and t

|5| $$\frac{\partial}{\partial t} u(x, t) = k \frac{\partial}{\partial x} \left[ f(u) \frac{\partial}{\partial x} u(x, t) \right]$$

where f(u) takes the form

|6| $$f(u) = u^s$$

The following functional transformation

|7| $$\bar{u} = F(u) = \int_0^u f(u)\, du$$

allows us to write |5| in the form

|8| $$\frac{\partial \bar{u}}{\partial t} = k\, H(\bar{u}) \frac{\partial^2 \bar{u}}{\partial x^2}$$

except for s =-1, with

|9| $$H(\bar{u}) = \left[ (s+1)\, \bar{u} \right]^{s/(s+1)}$$

Taking the infinitesimal group transformation

$$\hat{\bar{u}} = \bar{u} + \varepsilon B(u,x,t) + \theta(\varepsilon^2)$$

|10| $$\hat{t} = t + \varepsilon T(u,x,t) + \theta(\varepsilon^2)$$

$$\hat{x} = x + \varepsilon A(u,x,t) + \theta(\varepsilon^2)$$

and invoking the formal invariance of equation |8|

$$\frac{\partial\bar{u}}{\partial t} - k'\bar{u}^{s/(s+1)} \frac{\partial^2\bar{u}}{\partial x^2} = \frac{\partial\hat{\bar{u}}}{\partial\hat{t}} - k'\hat{\bar{u}}^{s/(s+1)} \frac{\partial^2\hat{\bar{u}}}{\partial\hat{x}^2} = 0$$

where $k' = k\,(s+1)^{s/(s+1)}$ we obtain

|11| $T = at + b \quad ; \quad A = cx + d \quad ; \quad B = (2c - a)\,u$

The non-essential parameter a in |11| can be eleminated to give

|12| $T = t + b' \quad ; \quad A = c'x + d' \quad ; \quad B = (2c' - 1)\,u$

These generators give the following group operator

|13| $\chi = (t + b')\frac{\partial}{\partial t} + (c'x + d')\frac{\partial}{\partial x} + (2c' - 1)\,u\frac{\partial}{\partial u}$

and the corresponding invariants

|14| $J_1 = \dfrac{x + d'/c'}{(t' + b')^{c'}} \quad ; \quad J_2 = \dfrac{u}{(t + b')^{(2c' - 1)}}$

We shall return to this problem later on in connection with self-similar methods, and invariant solutions will be obtained. But it is very interesting to consider the element of the group given by $b' = d' = 0$, $c' = 1$, the operator of which is

|15| $\chi = t\frac{\partial}{\partial t} + x\frac{\partial}{\partial x} + u\frac{\partial}{\partial u}$

Note that |15| generates a stretching (homogeneous) group because it has the form

$$\sum_{i=1}^{n-m} a^i x^i \frac{\partial}{\partial x^i} + \sum_{k=1}^{m} b^k u^k \frac{\partial}{\partial u^k}$$

where $a^i$ and $b^k$ are arbitrary constants.

This means that the concept of invariant solutions under a Lie group contains, in part, the concept of the self-similar solutions or automodel solutions, if they exist.

Ordinarily the concept of an automodel solution is presented connected to dimensional analysis, although here this concept loses its meaning. We are going to apply this self-similar method to a concrete example in paragraph IV.

c) Self-similar technique

The self-similar groups are one parameter groups, most of them stretching groups, which allow invariant solutions to be found for a system of non-linear partial differential equations. But the technique to obtain the invariants is much simpler.

In this case we take a transformation of the form

$$\hat{x}_i = a^{\alpha_i} x_i$$

|16|

$$\hat{u}_i = a^{\beta_i} u_i$$

Let L = 0 be again a system of partial differential equations. A self-similar group as |16| is applied to keep the system formally invariant and this invariance produces the relations involving the arbitrary parameters $\alpha_i$ and $\beta_i$ which determine a particular element of the group. The subsequent analysis of the transformations will give the corresponding invariants. From this point the same method described for infinitesimal groups is used to obtain the reduced system determining the invariant solutions.

We treat by this method the previous example, the non-linear heat diffusion equation, and some solutions will be given.

d) Example : heat equation

Taking the heat equation as |8|

|17| $$\frac{\partial \bar{u}}{\partial t} = k' \, \bar{u}^n \frac{\partial^2 \bar{u}}{\partial x^2} \qquad \text{with} \qquad n = \frac{s}{s+1}$$

we invoke formal invariance after application of a similarity group

$$\hat{\bar{u}} = a^{\alpha}\bar{u} \quad ; \quad \hat{t} = a^{\beta}t, \quad \hat{x} = a^{\gamma}x$$

and we obtain

$$\alpha = \frac{1}{n}(2\gamma - \beta)$$

with $\gamma$ and $\beta$ arbitrary constants.

For sake of simplicity one can take

$$\gamma = \omega\beta$$

to write

|18| $$\hat{\bar{u}} = a^{(2\omega-1)\beta/n}\bar{u} \quad ; \quad \hat{t} = a^{\beta}t \quad ; \quad \hat{x} = a^{\omega\beta}x$$

with invariants

$$|19| \qquad J_1 = x \left(\frac{t}{T}\right)^{-\omega} \qquad ; \qquad J_2 = \bar{u} \left(\frac{t}{T}\right)^{(1-2\omega)\ 1/n}$$

We dispose of an arbitrary parameter $\omega$. T is a characteristic time for the evolution of the system and the origin of time should for convenience be taken at t = T. Since for t = T, the invariants become

$$J_1 = x \qquad ; \qquad J_2 = \bar{u}$$

(of course a translation of time is always possible and we could introduce the time variable t-T).

In order to obtain invariant solutions, we assume

$$|20| \qquad J_2 = G(J_1)$$

and we substitute |19| in eq. |17| which becomes

$$|21| \qquad \frac{1}{n}(2\omega - 1)\, G(\eta) - \omega\eta \frac{dG}{d\eta} = Tk'\, G^n \frac{d^2G}{d\eta^2}$$

with

$$\eta = J_1 = x \left(\frac{t}{T}\right)^{-\omega}$$

We choose $\omega$ in such away that the left hand side of |21| multiplied by $G^{-n}$ is a total differential. It gives $\omega = (s + 2)^{-1}$. Then, obviously except for s = -2, eq. |18| can be written in the form

|22| $$\frac{d^2G}{d\eta^2} = k'' \frac{d}{d\eta}\left(\eta G^{1/(s+1)}\right)$$

where

|23| $$k'' = -\frac{(s+1)}{k'T\,(s+2)} = -\frac{(s+1)^{1/(s+1)}}{KT\,(s+2)}$$

The invariant solutions obeying the restriction

$$\left.\frac{dG}{d\eta}\right|_{\eta=0} = 0 \quad \text{with} \quad G\Big|_{\eta=0} \ \text{arbitrary}$$

is defined by the family of curves

$$G = (a\eta^2 + b)^c$$

with b arbitrary constant ($G|_{\eta=0} = b^c$), a and c defined by

$$a = -\frac{s(s+1)^{-s/(s+1)}}{2kT\,(s+2)} \quad ; \quad c = \frac{s+1}{s}$$

and taking the inverse transformations we obtain the following family of solutions to the non-linear partial differential heat equation[12]

$$u = \left[(s+1)\bar{u}\right]^{1/(s+1)}$$

|24| $$u = \left[\frac{Ax^2 + B\left(\frac{t}{T}\right)^{2/(s+2)}}{(t/T)}\right]^{1/s}$$

where B is arbitrary and A is given by

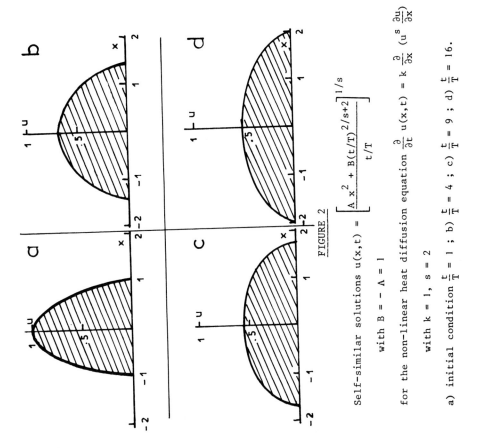

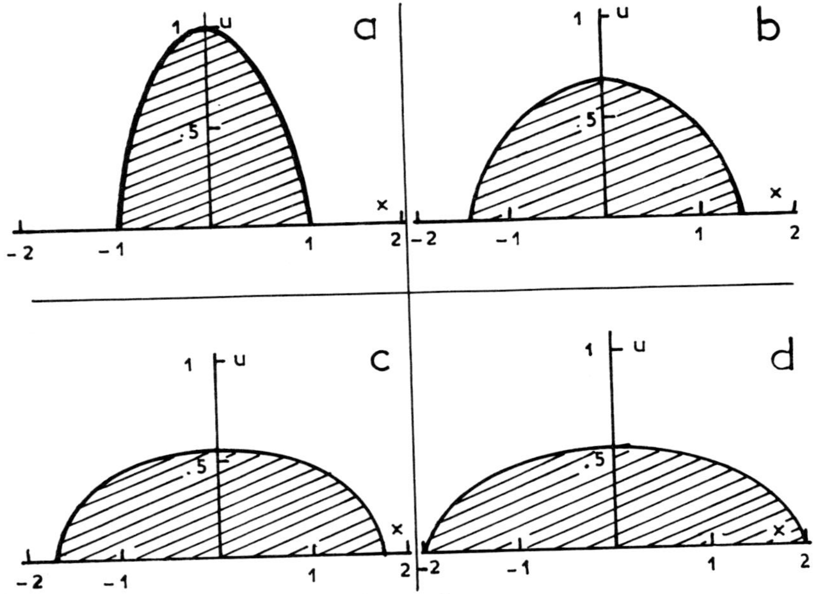


FIGURE 2

Self-similar solutions $u(x,t) = \left[\frac{A\ x^2 + B(t/T)^{2/s+2}}{t/T}\right]^{1/s}$

with $B = -A = 1$

for the non-linear heat diffusion equation $\frac{\partial}{\partial t}\ u(x,t) = k \frac{\partial}{\partial x}\ (u^s \frac{\partial u}{\partial x})$

with $k = 1$, $s = 2$

a) initial condition $\frac{t}{T} = 1$ ; b) $\frac{t}{T} = 4$ ; c) $\frac{t}{T} = 9$ ; d) $\frac{t}{T} = 16$.

$$A = -\frac{s}{2kT\ (s+2)}$$

In this case we have a number of interesting physical initial conditions for instance for all values of s with $s > 0$. Notice that the solution is truncated below the points where u cancels (fig. 2).

## III. PHASE SPACE FLUIDS

We are going now to consider some solutions obtained for those systems, in which we are mainly interested, namely the phase space fluids. We call phase space fluid, a N-body system described by a probability density obeying a continuity equation and other physical constraints defined on $R^{2n}$ (configuration and velocity spaces). In contradistinction some fluids can be described by the two first moments of the probability density which is a considerable simplification. This is the case for water, or non rarefied neutral gas.

### a) Three particular fluids

There are three particular cases which can be described mathematically by the same system of equations with non-linearity determined by an integral term.

The only interactions between particles considered in the system will be long ranges ; inelastic collisions and short range elastic collisions are neglected because they are exceptional processes, or because the evolution of the system is faster than the time between two close processes at a short range

$$\tau \text{ collisionless} \ll \frac{1}{\nu \text{ collision}}$$

where $\nu$ collision is the collision frequency and $\tau$ collisionless the characteristic evolution time of the system when only long range processes are considered.

These three collisionless phase space fluids are : two species plasmas, one species beams of charged particles and stellar systems.

A two species plasma is a system of two kinds of particles with opposite charge (ions and electrons) interacting accordingly to Coulomb's law.

A one species beam is a system of identical charged particles (ions or electrons) where the Coulomb forces will be only repulsive.

System constituted by N massive bodies (interstellar gas, set of stars or galaxies, etc...) interacting between them via the gravitational forces will be called stellar self gravitating systems.

These collisionless fluids will obey Liouville's equation with their corresponding electromagnetic or gravitational Hamiltonians. In the charged particle case the possibility of self-induced electromagnetic radiation fields must be taken into account due to the mobility of particles.

b) Mathematical description

Due to the impossibility of solving Liouville's equation in $R^{2n}$, simplifications must be introduced in the initial problem. For the above mentioned cases the BBGKY hierarchy[13] can be used with additional assumptions about the different correlation functions. Likewise we can follow the Rosenbluth and Rostoker method[14] to reach the same result as

- Vlasov + Maxwell equations -

In addition, for the plasma, the extrapolation to infinity for ionic masses is usually considered because they have very slow dynamics compared with electrons. In this way only an evolution equation for one species of particles remains, with ions appearing as a constant density in the divergence of the electric field equation (see Eq. |29|).

To sum up, the Vlasov approach is equivalent to take the limit of the mass and charge of electrons as going to zero, with the density of particles going to infinity such that $\frac{e}{m}$, ne and nm remain constant (e, m and n electronic charge, mass and density respectively). In this limit the plasma becomes a continous fluid.

This approximation can be visualized assuming that we "cut" the physical particles in two identical parts. Each new "particle" is again cut in two parts and so on, obtaining finally a "mash" which represents mathematically a continous fluid with loss of the grain character of the initial description.

The determining parameter of this situation is defined in the form

$$|25| \qquad g = \frac{1}{nL_D^3}$$

where $L_D$ is Debye's length or range of the screening effect. For a plasma

$$|26| \qquad L_D = \left(\frac{\varepsilon_0 \, kT}{4\pi \, ne^2}\right)^{1/2}$$

k being the Boltzmann constant.

Non-collisional cases correspond to

$$g \ll 1$$

whereas $g > 1$ indicates that individual effects are at least of the same order of magnitude than collective ones.

The equations obtained for plasma fluids are in the limit $g \to 0$

$$|27| \qquad \frac{\partial f}{\partial t} + \vec{v} \cdot \frac{\partial f}{\partial \vec{r}} + \vec{\gamma} \cdot \frac{\partial f}{\partial \vec{v}} = 0$$

$$|28| \qquad \vec{\gamma} = \frac{e}{m} (\vec{E} + \frac{1}{c} \vec{v} \times \vec{B})$$

$$|29| \qquad \operatorname{div} \vec{E} = 4\pi \, e \left[ \int_{-\infty}^{+\infty} f \, d^3\vec{v} - (N_0) \right] + 4\pi \, \rho_{ext}$$

$$|30| \qquad \operatorname{div} \vec{B} = 0$$

$$|31| \qquad \operatorname{rot} \vec{E} = - \frac{1}{c} \frac{\partial \vec{B}}{\partial t}$$

$$|32| \qquad \operatorname{rot} \vec{B} = \frac{4\pi}{c} \vec{J} + \frac{1}{c} \frac{\partial \vec{E}}{\partial t}$$

where $N_0$ is the constant density of ions, $\rho_{ext}$ corresponds to the external charges creating the external field, the integral term in eq. |29| is the mean field produced by all electrons in the plasma, and f the probability density in phase space, being a function of $\vec{r}$, $\vec{v}$ and t

$$f = f(\vec{r},\vec{v},t)$$

The mathematical description of a Beam is identical to the above one except in the equation |29| because $N_0$ is here equal to zero.

In these two cases we have written the most general system of equations. Neverthless, there exist interesting and practical limiting cases called "electrostatic" where the self-induced magnetic field due to mobility of particles is neglected. The equations are reduced to the Vlasov-Poisson system (moreover we have also taken the external magnetic field equal to zero)

|33| $$\frac{\partial f}{\partial t} + \vec{v}\cdot\frac{\partial f}{\partial \vec{r}} + \frac{e}{m}\vec{E}\cdot\frac{\partial f}{\partial \vec{v}} = 0$$

|34| $$\frac{\partial \vec{E}}{\partial \vec{r}} = 4\pi e\left[\int_{-\infty}^{+\infty} f\, d^3\vec{v} - (N_0)\right]$$

with $N_0 = 0$ for the beam case. The other Maxwell equations are automatically satisfied.

For a gravitational system the equations are formally identical to the beam case ones when the external forces are not considered. The only difference is the change of sign for the field divergence because the attractive character of interactions between particles. Equations |33| and |34| become

|35| $$\frac{\partial f}{\partial t} + \vec{v}\cdot\frac{\partial f}{\partial \vec{r}} + \vec{G}\cdot\frac{\partial f}{\partial \vec{v}} = 0$$

|36| $$\text{div}\,\vec{G} = -4\pi\, G_0\, m \int_{-\infty}^{\infty} f\, d^3\vec{v}$$

The techniques pointed out in the previous paragraph have made it possible to find some singular solutions for these three phase space fluids. These solutions exhibit difficulties on boundaries with, sometimes, particles located at infinity in configuration space and infinite kinetic energy. Nevertheless a physical meaning can be given to these analytical solutions (11). The used concept is called contamination and is based on the virtual separation of the particles in two populations, a central one (which has a physical meaning), and the external particles (which in fact we want suppress). The

The crucial point is that for systems with a sufficient degree of symmetry the interactions of the external particles on the central populations manifest itself only after a certain time in the worst case and sometimes never manifest, at least for some subsets of the central (physically meaningful) population. This concept has been quite useful especially in the beam case.

Applying the self-similar group techniques to equations |33| and |34|, we get the following invariants

$$|37| \qquad F = f\left(\frac{t}{T}\right)^{\alpha+1} \; ; \; \vec{\varepsilon} = \vec{E}\left(\frac{t}{T}\right)^{2-\alpha} \; ; \; \vec{\xi} = \vec{X}\left(\frac{t}{T}\right)^{-\alpha} \; ; \; \vec{\eta} = \vec{V}\left(\frac{t}{T}\right)^{1-\alpha}$$

with $\alpha$ a real arbitrary parameter and T the characteristic time in the evolution of the system. We obtain, sustituting |37| in |33| and |34| for a one dimensional system

$$|38| \qquad \eta \frac{\partial F}{\partial \xi} + \frac{e}{m} \varepsilon \frac{\partial F}{\partial \eta} - \frac{1}{T} \left\{ (\alpha-1)\eta \frac{\partial F}{\partial \eta} + \alpha \xi \frac{\partial F}{\partial \xi} + (\alpha+1) F \right\} = 0$$

$$|39| \qquad \frac{\partial \varepsilon}{\partial \xi} = 4\pi e \left[ \int_{-\infty}^{+\infty} F d\eta - N_0 \left(\frac{t}{T}\right)^2 \right]$$

Again, the time origin must be taken at t = T. Unfortunately, for $N_0 \neq 0$ the variable t is still present in the field divergence equation, which makes it impossible for the stretching groups to be used in the plasma case. But, for the gravitational and beam cases, $N_0 = 0$ invariant solutions can be obtained from eqs. |38| and |39|. In the one dimensional simple Water Bag model, open invariant solutions have been found(15) (e.g. solutions with particle at infinity in configuration space). However, as we have already pointed out, these solutions give information for all times on the behaviour of the central part of a physical system with closed initial conditions. The proportion described analytically depends on the initial dimensions of the system in phase space. These solutions are called "rod" solutions because of their form (see fig. 3).

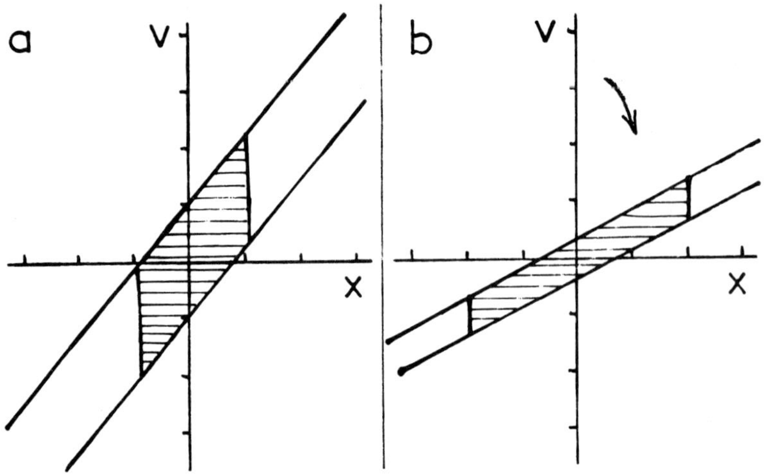


FIGURE 3

Self-similar solutions of a beam of identical changed particles in a single Water-Bag model. a) initial condition ; b) temporal evolution. The density decreases with time.

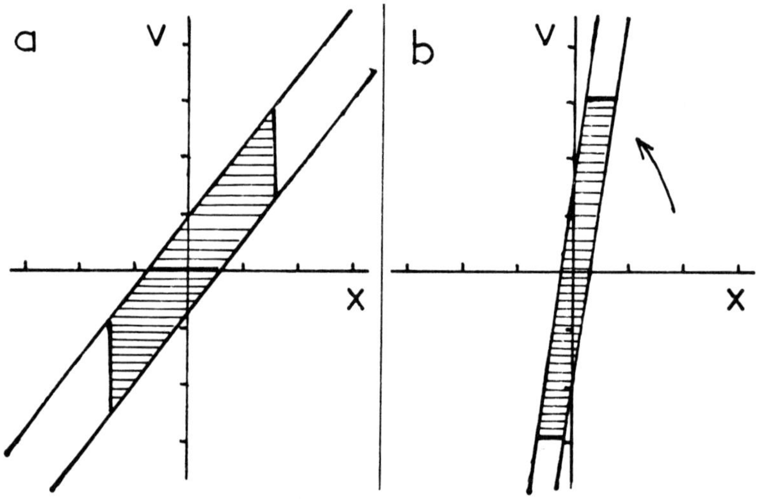


FIGURE 4

Collapse of the self-similar solutions of a gravitational problem in a single Water-Bag model. a) initial condition ; b) temporal evolution after a finite time.

In the one dimensional gravitational case we find solutions with the same structure, but their temporal evolution induces a genuine gravitational collapse, the energy being, after a finite time, concentrated in a very little region around the center of mass of the system. The model used is again a simple Water Bag model (see fig. 4).

For the non initiated reader let us remember the concept of the simple one dimensional Water Bag model. All particles are initially located with the same probability in a closed region of phase space ; that is to say the distribution function f is constant in this region and takes a zero value everywhere else. The Vlasov equation preserves this constant value of f for any time and consequently we have just to study the time evolution of the boundary curves which represent the upper, $V^+$, and lower, $V^-$, velocities in phase space. They obey the following equations

$$|40| \quad \frac{\partial}{\partial t} V^{\pm} + \frac{1}{2} \frac{\partial}{\partial x} (V^{\pm})^2 = E$$

$$|41| \quad \frac{\partial E}{\partial x} = A(V^+ - V^-) - (N_0)$$

A is the constant value of the distribution function. Obviously any distribution function in phase space can be approached by a multiple Water Bag model, but we are not going to study this generalisation.

For a three dimensional gravitational system with spherical symmetry the corresponding equations can be simplified to[24]

$$|42| \quad \frac{\partial F}{\partial t} + V_{//} \frac{\partial F}{\partial r} + \left( \frac{V_{\perp}^{2}}{r} - \frac{\partial \phi}{\partial r} \right) \frac{\partial F}{\partial V_{//}} - 2 \frac{V_{//} V_{\perp}^{2}}{r} \frac{\partial F}{\partial (V_{\perp}^{2})} = 0$$

$$|43| \quad \Delta \phi = 4\pi G_0 \rho$$

with $$\rho(t,r) = \pi \int_{-\infty}^{+\infty} dV_{//} \int_{0}^{\infty} d(V_{\perp}^2)\ F(t,r,V_{\perp}^2,V_{//})$$

|44| $$\frac{\partial\phi}{\partial r} = \frac{G_o}{r^2}\ m$$

$$m(r,t) = \pi \int_{0}^{r} d\,x\ 4\pi\ x^2 \int_{-\infty}^{+\infty} dV_{//} \int_{0}^{\infty} d(V_{\perp}^2)\ F(t,r,V_{\perp}^2,V_{//})$$

where $F(t,r,V_{//},V_{\perp}^2)$ is the distribution function of the system satisfying the Vlasov (Boltzmann) equation |42| with the independent variables : time t, radius r, radial velocity $V_{//}$ and the square of tangential velocity $V_{\perp}^2$. $\phi$ is the interaction potential in the system with $\rho(r,t)$ the particle density and $m(r,t)$ the mass located within a distance r at a time t. $G_o$ is the gravitational constant.

In this example we will apply the concept of self-similar invariance twice, up to the step where analytical results can be obtained. We hope to still retain some information about the non linear behaviour of the system.

At first, we "absorb" the time in the new variables and functions ; the invariants obtained are

|45|
$$H = F(t/T)^{2-3\alpha} \ ; \quad M = m(\tfrac{t}{T})^{3\alpha-1}$$

$$\xi = r(t/T)^{\alpha-1} \ ; \quad \eta = V_{//}(t/T)^{\alpha} \ ; \quad \lambda = V^{\ 2}(t/T)^{2\alpha}$$

which generate the following reduced equations

|46| $$\frac{(3\alpha-2)}{T}\ H + \frac{1}{T}\left\{(\alpha-1)\ \xi\,\frac{\partial H}{\partial \xi} + \alpha\,\eta\,\frac{\partial H}{\partial \eta} + 2\,\alpha\,\lambda\,\frac{\partial H}{\partial \lambda}\right\}$$

$$+ \eta\,\frac{\partial H}{\partial \xi} + \frac{1}{\xi}\left\{\lambda - G_o\,\frac{M}{\xi}\right\}\frac{\partial H}{\partial \eta} - 2\,\eta\,\frac{\lambda}{\xi}\,\frac{\partial H}{\partial \lambda} = 0$$

$$|47| \qquad \frac{dM}{d\xi} = 4\pi\xi^2 \int_{-\infty}^{+\infty} d\eta \int_0^{\infty} d\lambda \;\; H(\xi,\eta,\lambda)$$

with T the characteristic time in the system (initial time corresponds to t = T), and $\alpha$ an arbitrary constant.

Now we absorb the space variable $\xi$ through a new application of the self-similar group of transformations technique. In this process the reduced mass M of the system will become a constant. The corresponding invariants are here :

$$|48| \qquad \begin{aligned} D &= H(\xi/R)^3 \quad ; \quad \Omega = M(\xi/R)^{-3} \\ \omega &= \eta(\xi/R)^{-1} \quad ; \quad \phi = \lambda(\xi/R)^{-2} \end{aligned}$$

giving the equations

$$|49| \qquad D(\frac{R}{T} - 3\omega) + (\phi - \omega^2 + \omega\frac{R}{T} - G_o \frac{\Omega}{R}) \frac{\partial D}{\partial \omega}$$

$$+ (2\phi\frac{R}{T} - 4\phi\omega) \frac{\partial D}{\partial \phi} = 0$$

$$|50| \qquad \Omega = \frac{4}{3}\pi R^3 \int_{-\infty}^{+\infty} d\omega \int_0^{\infty} d\phi \;\; D(\omega,\phi)$$

In |48| and |49| R is the space scaling factor.

At this step the solutions of eqs. |49| and |50| can be calculated, and we can write

$$|51| \qquad D = q^{-3/4} \left[\frac{q + (p+\sqrt{A})^2}{q + (p-\sqrt{A})^2}\right]^{B/2\sqrt{A}} K\left[\frac{A-(p^2+q)}{q}\right]$$

where $K(\sigma)$ is an arbitrary function of its argument $\sigma$ and the new variables are in terms of the old ones

|52|

$$p = 4\omega - 2B$$

$$q = 4\phi$$

$$B = R/T$$

$$A = 4\left(\frac{R}{T}\right)^2 \left(1 - \frac{4}{3} J^2 T^2\right)$$

Notice that A is a parameter which depends on the Jean's frequency

|53| $$J = (4\pi G_0 \rho_0)^{1/2}$$

$\rho_0$ being the mean density in the system.

Coming back to the initial variables $r, t, V_{//}, V_\perp^2$ we notice that this solution corresponds to the expansion of a system with a density of particles uniform in space and varying as $\rho_0 (t/T)^{-2}$.

The nonlinearity of the problem subsits through the fact that only some values of A are permissible. These values are obtained by noticing that D must verify the Poisson equation |50|. In fact it will verify |50| provided the integral of the left hand side exists. The limiting values of A are $4B^2$ (with corresponds to a ballistic expansion with a negligible effect of the gravitational forces) and $A = \frac{4}{9} B^2$ (which corresponds to $J^2 T^2 = \frac{2}{3}$). This solution corresponds to the slowest self similar expansion (for a given initial density $\rho_0$).

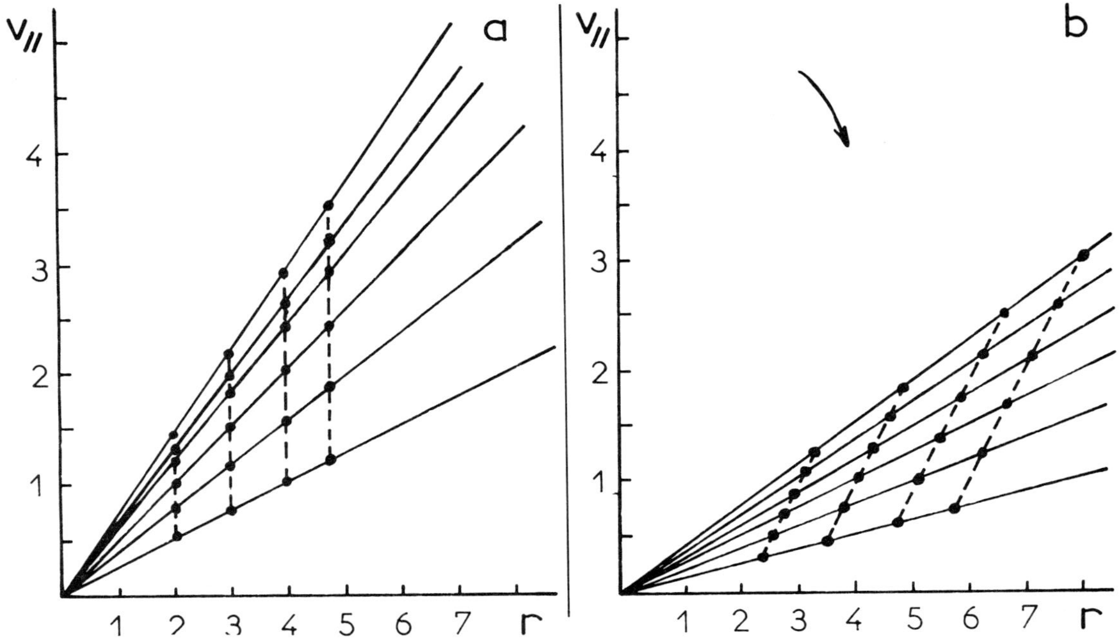


FIGURE 5

Phase plane diagram for the radial solution of the selfsimilar gravitating model with

J T = 3/4

a) initial condition $\frac{t}{T} = 1$. At this time $\frac{1}{4}\frac{R}{T} < V_{//} < \frac{3}{4}\frac{R}{T}$

b) temporal evolution for $\frac{t}{T} = 2$. Each point represents a given number of particles.

These solutions present the same difficulty as in the previously mentionned one-dimensional case, since they have particles at infinity in configuration space. We show in figure |5| the phase space structure of some solution of |51| - Those for which $V_{\perp}$ is always zero - We can appreciate their open structures. As in the one-dimensional case a central part describes an initial closed physical system. For this central part an analytical description is made possible by the introduction of the concept called "contamination effect".

We have given this rather lengty calculation to show how by successive applications of the self similar technique a very complex coupled non-linear system of equations can be brought to fully analytical analysis. Of course, in this process many interesting physical situations have been suppressed. Nevertheless we still recover interesting physical results on the expansion of homogeneous systems, the role of gravitational forces and their velocity structure.

In the next section we are going to present a very interesting application of self-similar group techniques where we shall not be interested in the invariant solutions but in the invariants themselves, because they give us enough information to construct a scaled system from an initial one. Notice that a self-similar transformation is often a stretching transformation, and this means a rescaling of the variables.

## IV. AN APPLICATION OF SELF-SIMILAR GROUPS : SCALING LAWS

The theory of models can be assimilated to the existence of some combinations of the physical parameters characterizing a system, which remain invariant under a transformation group. Consequently changes in the parameters of the model can be introduced without changing its physics.

In this sense, a self-similar group can define scaling laws for a model characterised by a system of equations, since their transformations are written in the form

|54| $$\hat{x}_i = a^{\alpha i} x_i$$

In principle the new equations obtained can be different from the initial ones, but, here, in order to deduce the scaling laws we will impose the formal invariance, their form remaining unchanged. It introduces a system of linear algebraic homogeneous equations for the different $\alpha_i$. If this system has solution in addition to the trivial solution $\alpha_i = 0$, we have a possible scaling law. In fact, the number of independent $\alpha_i$ is equal to the number of degrees of freedom, i.e. the maximum number of parameters in the model which can be changed independently. If only one $\alpha_i$ is arbitrary we can change one parameter ; for two $\alpha_i$ we can choose arbitrarily at most two parameters, although not all pairs are allowed.

Let us suppose that we know the system of equations which describes the whole physics of a Fusion Machine. Invariance under a self-similar group will give the group invariants, through which we can express any solution of the system as an arbitrary function. If this function remains unchanged the physics in the machine is also unchanged. This is possible if the parameters are scaled whilst preserving the group invariants. We have then constructed a "self-similar" or an automodel machine.

In this application there is a very important fact ; we are not obliged to know the explicit form of solutions. In this philosophy the measurement of one machine is automatically extrapolated to other machines and this "self-similar" behaviour defines a family or determined type of fusion machine for which prediction is possible. Now it will be very interesting to know if, by scaling, they can reach Lawson's criterion

|55| $$n\,\tau \geq 10^{14}$$

where n (in $cm^{-3}$) is the plasma density and $\tau$ (in s) the confinement time.

It must be pointed out that this method has a lot of limitations. Of course we do not claim to have solved the fusion problem because the whole physics of any fusion machine is not known at present. But it seems to us interesting to study some physical models of plasmas which are involved in part by Tokamaks and other experimental thermonuclear devices, in order to decide if they are interesting from this physical point of view.

On the other hand, the choice of a particular model (or set of equations) depends on which aspect of the problem is being considered. For a fusion machine three aspects can be considered

- the formation of a plasma from a neutral gas by dissociation and ionization

- the heating of this plasma

- the confinement of the fully ionized plasma.

The first phase implies phenomena involving quantum mechanics and atomic physics. The second phase leads to considering a set of charged particles in external and internal electromagnetic fields from a very low temperature to a high temperature (about $10^8$°K) in the frame of classical mechanics. The third phase involves the evolution of this population, under classical mechanics up to its disappearance by fusion processes.

The first two phases although important from a practical point of view can be forgotten for our derivation purposes and we shall suppose that the plasma is in the confinement step.

As has been stated in section III the Vlasov model is only valid for a plasma parameter g, (as defined in |25|), much smaller than 1, and if individual effects such as cyclotron radiation, bremstrahlung and collisions are neglected. This may be true for the confinement phase.

Let us consider three sucessive approaches to the model characterized by the Vlasov-Maxwell equations. As we introduce new physical phenomena into our system, the number of degrees of freedom will decrease. We will see that for Vlasov's model with self-consistant electromagnetic fields we have only one degree of freedom and the introduction of a new physical restriction makes it impossible to find a scaling law : for instance, collisions cannot be introduced. Self-similarity properties for more complicated models must be studied and the crucial question is the validity of these models. If we can be sure of the correctness of one of this model we must obtain the transformation group and see what freedom is left.

Here we start from eqs. |27| to |32| which describe the complete Vlasov-Maxwell model. A first approach consists in considering only an external magnetic field and moreover assuming that the plasma is at such low density that both the self-consistent electric and magnetic fields are negligible). The system of equations is reduced to the Vlasov equation only

|56| $$\frac{\partial f}{\partial t} + \vec{v} \cdot \frac{\partial f}{\partial \vec{r}} + (\vec{v} \times \vec{B}) \cdot \frac{\partial f}{\partial \vec{v}} = 0$$

where $\vec{B}$ is an external field which we can arbitrarily vary.

The self-similar group which leaves invariant eq. |56| is

|57| $$\hat{\vec{r}} = a^{\lambda} \vec{r} \quad ; \quad \hat{f} = a^{\alpha} f \quad ; \quad \hat{t} = a^{-\omega} t$$

$$\hat{\vec{v}} = a^{\lambda + \omega} \vec{v} \quad ; \quad \hat{\vec{B}} = a^{\omega} \vec{B}$$

the corresponding invariants being

|58| $$J_1 = B_i t \; ; \; J_2 = \frac{v_i}{r_j^{(1+c/b)}} \; ; \; J_3 = \frac{v_i^3 f}{r_j^{(1/b+3c/b+3)}}$$

$$J_4 = \frac{B_i}{r_j^{(c/b)}}$$

We note that $\lambda$, $\alpha$ and $\omega$ in |57| are three arbitrary parameters which give the possibility of obtaining a system with one, two, or three degrees of freedom. In the expression |58| i and j define any component of the vectors $\vec{B}$, $\vec{v}$ and $\vec{r}$. Due to the fact that only three variables are independant one of the four invariants will be an arbitrary function of the others. The invariants are expressed as functions of the variables characterizing the machine. We will, arbitrarily, select four quantities namely the radius of the machine "a", the confinement time $\tau$, the temperature of the plasma T and the plasma density n ; then expressions |58| can be written in the form

$$J_1 = B\tau, \quad J_2 = \frac{T}{a^2 B^2}, \quad J_3 = n\, a^{\phi}, \quad J_4 = B\, a^{\xi}$$

with $\phi$ and $\xi$ arbitrary parameters. The determined values of $\phi$ and $\xi$ define a family of self-similar machines. Choosing $J_1$ as a function of $J_2$, $J_3$ and $J_4$, we obtain

$$|59| \qquad B\tau = F\left(\frac{T}{a^2 B^2}, n\, a^{\phi}, B\, a^{\xi}\right)$$

F being an arbitrary function. If F is fixed, a particular type of machine is obtained. Now we can separate the devices with different degree of freedom

a) For $\phi = \xi = -\infty$ we have a machine with three degrees of freedom

$$|60| \qquad B\tau = F\left(\frac{T}{a^2 B^2}\right)$$

We consider the 4 parameters describing the machine and $\tau$ which indicates the performance. Among the 3 parameters T, a and B

2 can be chosen arbitrarily and moreover n remains always completely arbitrary. The last of the three parameters T, a and B and the performance $\tau$ are subsequently fixed. We can describe the system as possessing three degrees of freedom. This is obviously the most interesting and general scaling law for this model where self consistent effects are neglected.

b) For $\phi = -\infty$, $\xi > -\infty$ we have a device with two degrees of freedom

|61| $$B\tau = F\left(\frac{T}{a^2 B^2}, B a^s\right)$$ with s fixed arbitrarily

Here the choice of scaling factor for a variable, allows an arbitrary choice for two others, although not all pairs are permissible.

For each value of s a new self-similar family of machines is obtained.

c) For $\phi > -\infty$, $\xi = -\infty$ we obtain again a machine with two degrees of freedom

|62| $$B\tau = F\left(\frac{T}{a^2 B^2}, n a^r\right)$$

d) For $\phi > -\infty$, $\xi > -\infty$ devices with one degree of freedom only are obtained

|63| $$B\tau = F\left(\frac{T}{a^2 B^2}, n a^r, B a^s\right)$$

As has been said previously, if new physical restrictions are considered the arbitrary parameters $\phi$ and $\xi$ will be fixed and the number of degrees of freedom will decrease.

Let us consider now the electrostatic approach to this problem. The magnetic field is external and only the self-consistent electrostatic field is taken into account. From all the Maxwell equations only the Poisson equation is retained, a hypothesis obtained formally by letting c(velocity of light) go to infinity in the Maxwell equations |27-32|. The corresponding equations describing the system will be

|64| $$\frac{\partial f_i}{\partial t} + \vec{v} \cdot \frac{\partial f_i}{\partial \vec{r}} + (\vec{E} + \vec{v} \times \vec{B}) \cdot \frac{\partial f_i}{\partial \vec{v}} = 0$$

|65| $$\vec{\nabla} \cdot \vec{E} = \sum_i \int_{-\infty}^{+\infty} f_i \, d^3\vec{V}$$

where i indicates the different species in the plasma, that is electrons and ions. The self-similar group which leaves invariant |64| and |65| is defined by its transformations

|66| $$\hat{f} = a^{-(3\lambda+\omega)} f \; ; \; \hat{\vec{r}} = a^{\lambda}\vec{r} \; ; \; \hat{t} = a^{-\omega}\, t$$

$$\hat{\vec{V}} = a^{(\lambda+\omega)}\, \vec{V} \; ; \; \hat{\vec{B}} = a^{\omega}\vec{B} \; ; \; \hat{\vec{E}} = a^{(\lambda+2\omega)}\vec{E}$$

where we have only two arbitrary parameters and therefore at most two degrees of freedom.

The corresponding invariants can be contructed in a similar way to the previous case

$$J_1 = B\tau \; ; \; J_2 = \frac{T}{a^2 B^2} \; ; \; J_3 = \frac{n}{B^2} \; ; \; J_4 = B\, a^{\phi}$$

with $\phi$ an arbitrary parameter.

As for the previous case $\phi > -\infty$ give a self-similar set of devices with one degree of freedom and for $\phi = -\infty$ we have the interesting case of two degrees of freedom machines. The corresponding function is

|67| $$B\tau = F\left(\frac{T}{a^2 B^2}, \frac{n}{B^2}\right)$$

It is very easy to prove that if a collision term is added to eq. |64| to obtain Boltzman's equation the parameter $\phi$ is fixed with the value $\phi = 3/2$. In that case we can vary arbitrarily only one parameter, all the others and the performance being consequently deduced. We obtain

|68| $$B\tau = F\left(\frac{T}{a^2 B^2}, \frac{n}{B^2}, a^{3/2}B\right)$$

To finish, let us apply the self-similarity techniques to the whole Vlasov-Maxwell system defined by the eqs.

$$\frac{\partial f_i}{\partial t} + \vec{V} \cdot \frac{\partial f_i}{\partial \vec{r}} + (\vec{E} + \vec{V} \times \vec{B}) \frac{\partial f_i}{\partial \vec{V}} = 0$$

$$\text{div } \vec{E} = \sum_i \int_{-\infty}^{+\infty} f_i \, d^3\vec{V}$$

$$\text{div } \vec{B} = 0$$

$$\text{rot } \vec{E} = -1/C \frac{\partial \vec{B}}{\partial t}$$

$$\text{rot } \vec{B} = \frac{4\pi}{C} \vec{J} + \frac{1}{C} \frac{\partial \vec{E}}{\partial t}$$

which remain invariant under the group characterized by the following transformations

$$\hat{f}_i = a^{2\omega} f_i \; ; \; \hat{\vec{r}} = a^{-\omega} \vec{r} \; ; \; \hat{t} = a^{-\omega} t \; ;$$

$$\hat{\vec{V}} = \vec{V} \; ; \; \hat{\vec{B}} = a^{\omega} \vec{B} \; ; \; \hat{\vec{E}} = a^{\omega} \vec{E}$$

Here we have only one degree of freedom and the corresponding family of self-similar machines will be defined by

|69| $$B\tau = F\left(\frac{T}{a^2 B^2} , \frac{n}{B^2} , B\, a\right)$$

which derive from the invariants

|70| $$J_1 = B_K \tau \; ; \; J_2 = V_K \; ; \; J_3 = \frac{f_i}{B_K^{\,2}} \; ; \; J_4 = B_K\, r_j$$

As before K and j mean that the invariant is defined for any component of the vectors $\vec{B}$, $\vec{r}$ and $\vec{V}$.

It is not possible now to consider new physical restrictions to this problem, because of the incompatibility of the homogeneous determinant system. For example it is impossible to get scaling laws for a fully electromagnetic plasma including effects of collisions. As mentionned in |69| using only the electrostatic approximation allows the inclusion of collisions. |68| and |69| are the two most complete scaling laws for plasma.

Before commenting the previous results, we must say that a similar analysis has been done by Connor and Taylor(17) from a different point of view, and for particular cases contained here since a slightly more general treatement has been used here. We will consider only scaling laws as given by |69| and we note that the temperature T is an invariant. This means that no scaling law exist for machines at different temperatures, which turn out to be the most difficult thing to extrapolate physicaly. As soon as it is possible to obtain the correct temperature we can find interesting similarity laws on the influences of "B", "n" and "a" on $\tau$.

But there is one interesting idea that can be developed. Due to the fact that each function F defines a family of self-similar machines we can look at the effect on the physical variables in the machine when one of these variables is changed arbitrarily and the others subsequently deduced.

Each machine will now generate a one parameter family (i.e. having one degree of freedom) defined by a particular function F with the same plasma dynamics. The interesting point is that, for this family, the fusion performances will be different and also the technology. Fusion performances are measured by the product $n\tau$ of the density with the confinement time. Let us consider the $n\tau$ diagram in Figure 6 and let us draw the curve corresponding to the family generated by a given device. Since logarithmic scales are used, Lawson's criterion

|71| $$n\tau = 10^{14}$$

is represented by a straight line. The elimination of B between the invariants $J_1$ and $J_3$ in |70| gives

$$n\tau^2 = \text{const}$$

for the equation in n, $\tau$ of the family. Consequently, from a given machine we can deduce a family with better fusion performances. To increase the product $n\tau$ by a factor $\alpha$ we can

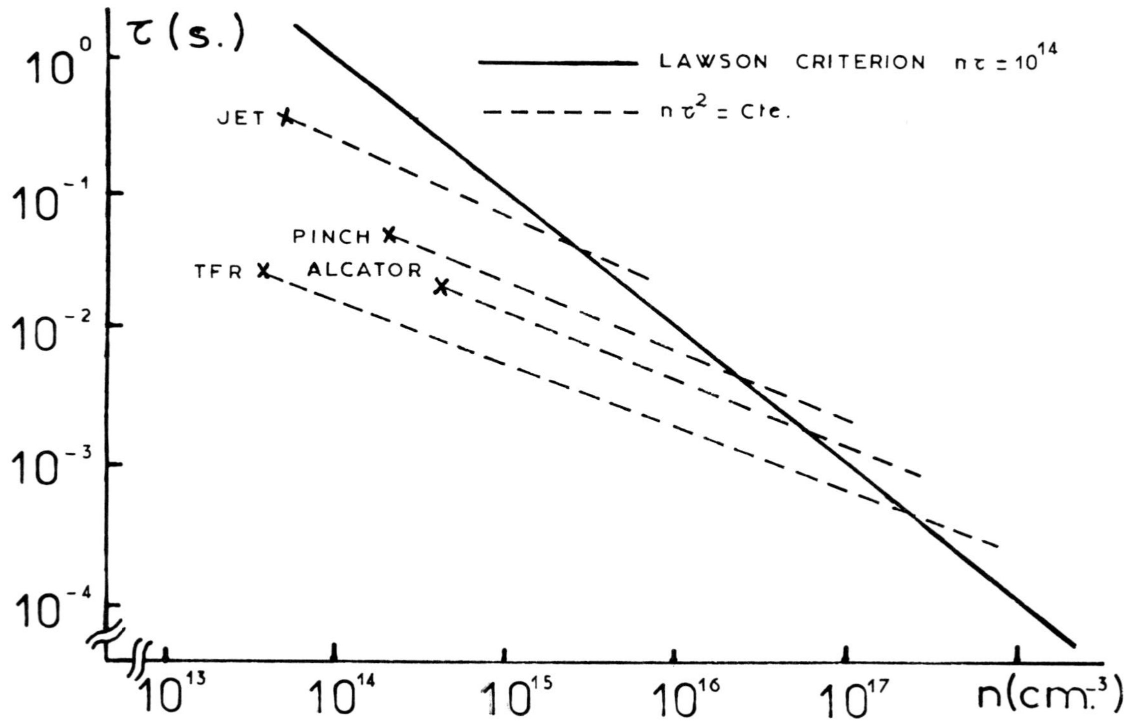


FIGURE 6

The broken lines show the different families of self-similar machines deduced from the Vlasov-Maxwell model for existing Tokamaks, in the density-time of confinement plane.

For each family, since $B\tau$ is an invariant the scaling factor for the two magnetic fields is given by the inverse ratio of the corresponding confinement times.

i) increase the magnetic field by the same factor $\alpha$

ii) decrease the characteristic size of the system by this factor $\alpha$

iii) increase the density by $\alpha^2$

The problem is now technological. Is it easier to build smaller machines with increased magnetic field ?. To specify the problem we look at different machines and extrapolate them to the Lawson criterion. TFR implies an increase of the magnetic field by more than two orders of magnitude (and the subsequent decrease in size). This gives a ridiculous machine of 1 cm, with a magnetic field of $5 \times 10^6$ gauss !

"Alcator" extrapolation is less unphysical. The ratio $\alpha$ to bring it on the Lawson curve is 12, implying a magnetic field of $10^6$ gauss with a torus radious of 5 cm. Nevertheless such a machine would be still quite unbalanced and impossible to build.

Thing change with JET family. Starting from a projected $\tau = 0.35$ s. and $n = 5 \times 10^{13}$ particles per $cm^3$ (and consequently $n\,\tau = 1.5 \times 10^{13}$) we will increase $n\,\tau$ to $10^{14}$ with a factor of 6 on the magnetic field (up to $18 \times 10^4$ gauss) and a decrease in size from 3 to 0.5 m. This clearly points out the advantage of a technological solution having an intense magnetic field and suggest that there is an interest in smaller machines.

We can suggest the following ideas :

- to test the scaling laws by building a set of small machines as self-similar as possible,

- to look at the family generated by each magnetically confined machine, performances of which in terms of $n\,\tau$ will always increase for smaller machines with a more intense magnetic field. It must be said that this statement will be considered under the scaling law philosophy and does not mean, of course, that smaller machines, all other parameters remaining unchanged, have better performances, but that we can exchange size against magnetic field and still obtain better performance.

This reinforces the interest in developping high magnetic field technology which will allow a substantial decrease in size (and consequently in price) of the machines. Connor and Taylor[18] have arrived at similar conclusions. Of course all the physics taking place in a Tokamak can hardly be brought down to the Vlasov-Maxwell model. Nevertheless this last model describes certainly very well the dynamics of the particles.

## V. TRANSFORMATION GROUPS OF PARTIAL INVARIANCE

### a) Transformations in time-phase space

As we have mentionned above, the formal invariance of a system of differential equations under a group of transformations presents two main difficulties : not any invariant solution is physical and solutions can be found only for a subset of initial or boundary conditions. In agreement with our philosophy we present here a new group of transformations which allows us to find solutions in a simpler way without reducing the number of independant variables. The invariance of the system of equations is only required "in part" and source terms can appear in the transformed system. Due to this fact groups of transformations like these will be called "GROUPS OF PARTIAL INVARIANCE". The newness here consists in a rescaling of the time through a function of t.

For a phase space system the following transformations are defined

|72|
$$\theta = \int_0^t \frac{dt}{C^2(t)}$$
$$\vec{\xi} = A(t)\,\vec{r}$$
$$\vec{\eta} = B(t)\,\vec{r} + D(t)\,\vec{v}$$

This transformation is somewhat connected to transformation introduced by Courant and Snyder(19) and rediscovered by Lewis(20). Here we show how to use it to solve some problems.

In order to keep invariant a system of equations as much as possible, the transformations defined by |72| are required to verify

$$d\vec{r}\, d\vec{v} = d\vec{\xi}\, d\vec{\eta} \qquad |73|$$

that is to say, the element of volume in phase space must be formally invariant. This means that the Jacobian of the transformations must be one. On the other hand, if the system derives from a Hamiltonian we impose two supplementary conditions

$$\frac{d\vec{\xi}}{d\theta} = \vec{\eta}, \qquad \frac{d\vec{\eta}}{d\theta} = \vec{\varepsilon} \qquad |74|$$

where $\vec{\eta}$ and $\vec{\varepsilon}$ are respectively the new velocity and field. Notice that we do not impose the strict formal invariance of Hamilton's equations.

The restriction |73| and |74| imply

$$A(t)\, D(t) = 1 \;;\quad C^2(t)\,\frac{dA}{dt} = B(t) \;;\quad A(t)\, C^2(t) = D(t)$$

then |72| becomes

$$\frac{d\theta}{dt} = \frac{1}{C^2(t)}, \qquad \vec{\xi} = \frac{1}{C(t)}\,\vec{r}, \qquad \vec{\eta} = C(t)\,\vec{v} - \frac{dC}{dt}\,\vec{r} \qquad |75|$$

It can be easily proved[21] that this set of transformation obeys the law of group and generates a continuous Lie group, C(t) being an arbitrary function.

These transformations can be generalized to the canonical variables in analytical mechanics with a new parametrization of the time[22] by writing

$$\frac{d\theta}{dt} = \frac{1}{C^2(t)}$$

|76| $$Q_i = \frac{1}{C(t)} q_i$$

$$P_i = C(t)\, p_i - \frac{dC}{dt} q_i$$

and it can be immediately verified that they keep formally invariant the Poisson brackets

$$\{\hat{H},\hat{F}\}_{PQ} = \frac{\partial \hat{H}}{\partial P_k}\frac{\partial \hat{F}}{\partial Q_k} - \frac{\partial H}{\partial Q_k}\frac{\partial F}{\partial P_k} =$$

$$= \frac{\partial H}{\partial p_r}\frac{\partial F}{\partial q_r} - \frac{\partial H}{\partial q_r}\frac{\partial F}{\partial p_r} = \{H,F\}_{pq}$$

where $\hat{H}(Q_r,P_r,\theta) = g\, H(q_r,p_r,t)$ is the transformed Hamiltonian (g being any element of the group), and F is any function of the independent variables. The Hamiltonian remain strictly invariant only for the trivial case C(t) = 1 (i.e. for the neutral element of the transformation group).

Nevertheless, we can define a new Hamiltonian $\mathcal{H}$ with ($\mathcal{H} \neq gH$) so that a system of Hamilton's equations can be written

|77| $$\overset{\circ}{Q}_r = \frac{\partial \mathcal{H}}{\partial P_r} \qquad \overset{\circ}{P}_r = \frac{\partial \mathcal{H}}{\partial Q_r}$$

It contains new interactions. Some can be traced back to the physical interactions, others entirely new, are connected to the change of phase space. In this sense the equations derived through the transformations |75| represent a new description of a physical problem with a "renormalization" of the forces.

An interesting fact arises from this description. If the function C(t), characterizing the transformation group, increases faster than $t^{1/2}$ when $t \to \infty$, the transformed time $\theta$ goes to a finite value. In this case we will say that the time is renormalized in the transformed system, which makes possible sometimes drastic simplifications for numerical calculations.

There exists a very frequent limiting case obtained for $C(t) = (1 + \Omega t)^{1/2}$ with $\Omega$ an arbitrary constant. This particular function generates a logarithmic compression of the initial time. For a form of C(t) with a leading term of degree smaller than one half the renormalization cannot be obtained and this type of transformation group becomes usually analytically less interesting. (But still interesting from a numerical point of view).

Another advantage of this method is the possibility to conter balance the time dependence of forces by a proper choice of C(t) and to obtain time independent Hamiltonian or at least one evolving in a slower way compare to its previous behaviour. We take advantage of this fact to solve this "new" problem.

Let us finish with a simple comment ; if the time renormalization is possible (without introducing infinities on the forces) we automatically can obtain informations on the temporal asymptotic limit for all initial conditions, without explicitly solving the equations. See paragraph below.

### b) Example of a non-linear harmonic oscillator

These groups of partial invariance have useful applications not only for the phase space fluids but for those systems in which a renormalization of time or forces can be obtained. We are going to comment here some interesting results and develop two particular examples ; namely the non-linear harmonic oscillator and the non-linear heat diffusion equation.

Let us start with the harmonic oscillator. Solutions with time dependent frequency have been obtained[23] for the linear cases characterized by the equation.

$$\frac{d^2 X(t)}{dt^2} + \omega^2(t)\; X(t) = 0$$

with frequencies expressed by

|78| $$\omega^2(t) = \frac{\omega_0^{\;2}}{(1+\Omega t)^{4\alpha}} \quad ; \quad \alpha > 0$$

$\omega_0$ and $\Omega$ are two constants.

Choosing for C(t) the following form

|79| $$C(t) = (1+\Omega t)^{\beta}$$

$$\text{with}\quad \begin{cases} \beta = \alpha & \text{for } 0 < \alpha < 1/2 \\ \beta = 1/2 & \text{for } 1/2 < \alpha < 1 \\ \beta = 1 & \text{for } 1 < \alpha \end{cases}$$

We obtain a time compression for $\alpha < 1/2$, a logarithmic compression for $1/2 < \alpha < 1$ and a time renormalization for $1 < \alpha$.

In a similar way it is possible to study the non-linear oscillator defined by

$$|80| \qquad \overset{\circ\circ}{X} + \omega^2(t) \left\{ X + k\, X^3 \right\} = 0$$

where k is an arbitrary constant, $\omega(t)$ being a real function of time. Taking for $\omega(t)$ and $C(t)$ the same expressions |78| and |79| the following transformed equation is derived from |80| by application of the transformations |75|

$$|81| \qquad \frac{d^2\xi}{d\theta^2} + \left[ \beta(\beta - 1)\, \Omega^2\, (1 + \Omega t)^{4\beta - 2} + \omega_0^{\,2}(1 + \Omega t)^{4(\beta - \alpha)} \right] \xi$$

$$+ k\, \omega_0^{\,2}(1 + \Omega t)^{6\beta - 4\alpha} \xi^3 = 0$$

This equation can be interpreted as a non-linear oscillator with a first linear term $\bar{\omega}^2\, \xi$ with $\bar{\omega}^2$ given by

$$\bar{\omega}^2(t) = \omega_0^{\,2}(1 + \Omega t)^{4(\beta - \alpha)}$$

completed by a time dependent linear force

$$- C^3 \frac{d^2 C}{dt^2}\, \xi = -\beta\, (\beta - 1)\, \Omega^2\, (1 + \Omega t)^{4\beta - 2}\, \xi$$

which will be called "transformation field" because it is a virtual field created by the transformation |75|, and finally the last term in the left hand side of |81| which is the non-linear term. The value of $\beta$ must be chosen in order to obtain an equation without infinite terms in the fields and to renormalize the time to simplify the numerical treatment. This last requirement implies a value of $\beta$ as high as possible within the limit of the first requirement which must be fulfilled first. With this "strategy" table I shows the different choices of $\beta$ when $\alpha$ runs from zero to infinity.

TABLE I :

| $\alpha$ | $\beta$ | $C^6(t)\omega^2(t)$ | $C^4(t)\omega^2(t)$ | $-C^3(t)\ddot{C}(t)\,\xi$ | $\Omega\,\theta$ |
|---|---|---|---|---|---|
| $0<\alpha<3/4$ | $\frac{2}{3}\alpha$ | 1 | $(1+\Omega t)^{-\frac{4}{3}\alpha}$ | $-\frac{2}{9}\alpha(2\alpha-3)\Omega^2(1+\Omega t)^{\frac{8\alpha-6}{3}}\,\xi$ | $\dfrac{(1+\Omega t)^{1-\frac{4}{3}\alpha}-1}{1-\frac{4}{3}\alpha}$ |
| $\alpha = 3/4$ | 1/2 | 1 | $(1+\Omega t)^{-1}$ | $\frac{1}{4}\,\Omega^2\,\xi$ | $\log(1+\Omega t)$ |
| $3/4<\alpha<3/2$ | 1/2 | $(1+\Omega t)^{3-4\alpha}$ | $(1+\Omega t)^{2-4\alpha}$ | $\frac{1}{4}\,\Omega^2\,\xi$ | $\log(1+\Omega t)$ |
| $\alpha = 3/2$ | 1 | 1 | $(1+\Omega t)^{-2}$ | 0 | $\Omega\theta = \dfrac{\Omega t}{1+\Omega t}$ if $t\to\infty$ |
| $\alpha > 3/2$ | 1 | $(1+\Omega t)^{6-4\alpha}$ | $(1+\Omega t)^{4(1-\alpha)}$ | 0 | $\Omega\theta \to 1$ |

We note here some essential differences with the linear case (see (23)) beginning by the different values of parameter $\alpha$ for which a renormalization of time is obtained.

We distinguish the following cases :

A) $0<\alpha<3/4$. Due to the choice of $\beta$, the term $C^6(t)\,\omega^2(t)$ becomes constant while the linear force $C^4\,\omega^2\,\xi$ is always smaller than 1 and decreases to zero with $\theta\to\infty$ together with the transformation field $-C^3\,\ddot{C}\,\xi$. In the linear case for $0<\alpha<1/2$ the physical force

$\omega^2\ C^4$ remains constant which indicates a solution x(t) with infinite amplitude when t goes to infinity. However in the non-linear case the term $C^6\ \omega^2$ is dominant over the linear one.

Consequently we write the equation which defines the limiting trajectories of the particules when t and $\theta \to \infty$

$$|82| \qquad \frac{d^2\xi}{d\theta^2} + k\,\omega_0^{\,2}\,\xi^3 = 0$$

and the asymptotic trajectories (limit cycles in the $\xi\ \frac{d\xi}{d\theta}$ plane) can be computed analytically.

B) $3/4 < \alpha < 3/2$. A renormalized time could have been obtained if we had taken $\beta = \frac{2}{3}\alpha$. Unfortunately the transformation field $-\ C^3\ \overset{\circ\circ}{C}\ \xi$ would be divergent with the subsequent difficulty of computation. Consequently we take the value $\beta = 1/2$ and obtain a logarithmic compression of time.

Now the leading term is the transformation field and the new asymptotic equation is simply

$$\frac{d^2\xi}{d\theta^2} = \frac{1}{4}\,\Omega^2\,\xi$$

C) $\alpha > 3/2$. It is now possible to obtain a time renormalization (i.e. $\Omega\theta \to 1$ when $t \to \infty$) by taking $\beta = 1$. The particles go to limit points $\xi_1$ in the $\xi$-space. Their asymptotic behaviour is directly calculated from the transformation itself

$$X_{limit} = (1 + \Omega t)\ \xi_1$$

This aspect of the problem shows the advantages of the method since now it is only necessary to solve the equation

$$|83| \qquad \frac{d^2\xi}{d\theta^2} + (1-\Omega\theta)^{4(\alpha-1)}\,\xi + (1-\Omega\theta)^{4\,\alpha-6}\,\xi^3 = 0$$

from $\theta = 0$ to $\theta = \Omega^{-1}$

to obtain asymptotic solutions. In this interval the eq. |83| does not present any infinite term, and then there will be no difficulty in the numerical calculation.

c) Example of the non-linear heat equation

Evidently, the main important feature presented by this type of transformations is the possibility of finding asymptotic solutions to a physical problem, even, if the time renormalization is not possible.

We shall return now to the non-linear heat diffusion equation defined previously

$$\frac{\partial u}{\partial t} = k \frac{\partial}{\partial x}\left(u^s \frac{\partial u}{\partial x}\right)$$

The transformation groups of partial invariance will be applied to obtain a transformed equation. Let us assume a transformation of the form :

$$\theta = \theta(t)$$

$$\xi = x.C(t)$$

$$\phi(\theta,\xi)\, B(t) = u(x,t)$$

where B(t), C(t) and θ(t) are in principle arbitrary functions. The heat equation becomes

$$|85| \qquad \frac{dB}{dt}\,\phi + B\,\frac{d\theta}{dt}\frac{\partial\phi}{\partial\theta} + \frac{dC}{dt}\frac{B}{C}\,\xi\,\frac{\partial\phi}{\partial\xi} = k\,B^{s+1}\,C^{2}\frac{\partial}{\partial\xi}\left(\phi^{s}\,\frac{\partial\phi}{\partial\xi}\right)$$

Particularizing $\mathring{B} = \beta\, B\mathring{\theta}$ and $\mathring{C} = \gamma C\mathring{\theta}$, with C = B = 1 for t = θ = 0 in order to obtain the same initial conditions in the two spaces, we get

$$B = \exp\beta\,\theta \quad ; \quad C = \exp\gamma\,\theta$$

Then, after division by $B\,\mathring{\theta}$, the equation |85| becomes

$$|86| \qquad \beta\,\phi + \gamma\,\xi\,\frac{\partial\phi}{\partial\xi} + \frac{\partial\phi}{\partial\theta} = k\,\frac{\exp\,(s\beta+2\gamma)\theta}{\mathring{\theta}}\,\frac{\partial}{\partial\xi}\left(\phi^{s}\,\frac{\partial\phi}{\partial\xi}\right)$$

Choosing the function θ in such a way that

$$|87| \qquad \frac{d\theta}{dt} = \exp\,(s\beta + 2\gamma)\theta$$

the expression of t as a function of θ is

$$t = \frac{1}{s\beta + 2\gamma}\left[1 - \exp\left[-(s\beta + 2\gamma)\,\theta\,\right]\right]$$

A compression of time is obtained when we take

$$2\gamma + s\beta = -\mu \quad \text{with } \mu > 0$$

which gives the usual logarithmic compression

|88| $$\mu\theta = \lg(1 + \mu t)$$

Now we ask for conservation of the heat flow in this transformed space. Integrating the function u(x,t) over the whole x space, we obtain

$$\int u \, dx = \int \phi \, d\xi \, \frac{B(t)}{C(t)} = \exp(\beta - \gamma)\,\theta \int \phi \, d\xi$$

If we want the different physical aspects to be the same in the two spaces, $\phi$ must be taken equal to zero as $\xi$ goes to infinity and imposing

$$Q = \int u \, dx = \int \phi \, d\xi = \overline{Q}$$

(Q being the quantity of heat), we must take

$$\beta = \gamma = - \frac{\mu}{s + 2}$$

in order to obtain $\frac{\partial \overline{Q}}{\partial \theta} = 0$, as it can be easily seen by integration over the whole $\xi$ space of equation |86| with the condition |87|. Then the non-linear heat diffusion equation is written

|89| $$- \frac{\mu}{s + 2} \left(\phi + \xi \frac{\partial \phi}{\partial \xi}\right) + \frac{\partial \phi}{\partial \theta} = k \frac{\partial}{\partial \xi} \left(\phi^s \frac{\partial \phi}{\partial \xi}\right)$$

and the transformations of partial invariance become

|90| $$\mu\,\theta = \lg(1 + \mu t)$$

$$x = \xi(1 + \xi t)^{1/s+2}$$

$$u(x,t) = \phi(\xi,\theta)\;(1 + \mu t)^{-1/s+2}$$

Let us study now the stationary solutions of eq. |89| in order to compare with the solutions obtained through the self similar groups. Imposing $\frac{\partial\phi}{\partial\theta} = 0$ and integrating |89|, we obtain

$$-\frac{\mu}{s+2}\,\xi\,\phi = k\phi^{s}\,\frac{\partial\phi}{\partial\xi}$$

As it has been done in the self-similar case, we take only those solutions with a maximum for $x = \xi = 0$ and the solutions obtained are written

$$\phi = \left(K - \frac{\mu s}{2(s+2)}\,\xi^{2}\right)^{1/s}$$

where K is a positive constant and k has been taken equal to 1. We note (as said previously) that the stationary solutions of |89| are identical to self-similar solutions obtained from |21| when $\omega$ is taken equal to $1/s+2$ ($\omega$ being the arbitrary parameter in self-similar transformations), and $1 + \mu t$ is replaced by $t/T$.

# C O N C L U S I O N

We have presented in this paper two kinds of philosophy on the use of group theory in Mathematical Physics, in view of getting solutions for non-linear partial derivative equations.

On the first philosophy we use group of transformations leaving the equations strictly invariant. Their use is limited by the following reasons.

→ They treat restricted initial or boundary conditions and it is difficult to say if slight change in these conditions will or will not modify considerably the structure of the solutions

→ If we use Lie group the equations to be solved are as difficult (sometimes more) than the initial equations

Among the possible easy to sort out transformations we must distinguish the stretching (self similar) ones. They are indeed very easy to obtain and sometimes give self similar solutions of some physical interest.

To balance these difficulties we notice

→ That for some problems the difficulties of the self similar solutions (usually divergence of some physical quantity at infinity) can be taken care of through the concept of contamination where we remark that the central part of a system with a sufficient degree of symmetry will not be immediately influenced by external parts (and sometimes will never be : this is the important contamination concept).

→ That self similar group techniques are quite interesting to search out the number of degree of freedom in the scaling of a system, and also to see, among many parameters, how many may be chosen independently and what subsets are permissible.

In the second philosophy we drop the strict invariance concept for the partial invariance one. Usually we change a problem into a problem of similar physics but where new terms make the system easier to solve (and by easier we include also a numerical simplification).

→ The time compression or time renormalization is the most important of the new concepts. Sometimes we can accelerate the time without introducing infinities in the forces or in the different terms of the equation. Then the usefulness is basically numerical. In the heat diffusion equations (linear and non-linear) a time logarithmic compression is quite obvious.

Of course, a time renormalization is still more interesting since it gives directly the asymptotic solutions. In both cases (compression or renormalization) the transformation is quite useful to compute the asymptotic form.

The field of applications are quite numerous. Here we have presented mostly applications to the evolution of phase space fluids (plasma, beams, gravitationnal gas where, incidently, plasma are more difficult to treat than the two others phase space fluids). Equations in configuration space can also be transformed. Problems involving diffusion equations and Schroedinger equations have been solved by the method(23). It will be interesting to see if hydrodynamics -which already make good use of self similar solutions- is also an interesting field for the application of these new techniques.

Last -but not least- the inclusion of these transformations within a numerical scheme is certainly an interesting application with its intriguing merging between analytical and numerical methods.

# BIBLIOGRAPHY

1 G. Kalman ; Annals of Physics 10, 1, (1960)

2 R. Balescu : "Statistical mechanics of charged particles". Wiley and Sons (London 1963)

3 D. Anderson, A. Bondeson, M. Lisak, R. Nakach and H. Wilhelmsson : EUR-CEA. FC 925 (October 1977). To be published in Physica Scripta.

4 F. Grant and M.R. Feix : Physics of Fluids 10, 696 (1967)

5 H.L. Berk and K.V. Roberts : Physics of Fluids 10, 1595 (1967)

6 S. Lie : "Theorie der transformations gruppen" I, II, III. Ch elsea Publ. Company. (New-York 1970)

7 L.V. Ovsjannikov : "Group properties of differential equations" Translated by G. Bluman (1967), Cal. Inst. of Tech.

8 W.F. Ames : "Non-linear partial differential equations in engineering", Vol II, Academic Press (1972)

9 H.S. Woodard : "Similarity solutions for partial differential equations generated by finite and infinitesimal group". PHD Thesis University of Iowa (1971)

10 G.W. Bluman and J.D. Cole : "Similarity methods for differential equations". Appl. Math. Sci. 13 Springer Verlag, N.Y. Heidelberg Berlin

11 J.R. Burgan, J. Gutiérrez, E. Fijalkow, M. Navet and M.R. Feix Journal de Physique Lettres 38 161 (1977)

12 W.F. Ames : "Non-linear partial differential equations in engineering". Vol I page 153 Academic Press (1965). The solutions obtained in this reference are found with particular auxilary conditions

13 N.N. Bogoliubov : "Studies in Statistical Mechanics" (Translated by E.K. Gora), Part A, Vol I. North Holland Publishing Co. (Amsterdam 1962)

14 N. Rostoker and M.N. Rosenbluth : Physics of Fluids 3, 1 (1960)

15 J.R. Burgan, J. Gutiérrez, E. Fijalkow, M. Navet and M.R. Feix : J. Plasma Physics 19, 135 (1978)

16 A. Munier, M.R. Feix, E. Fijalkow, J.R. Burgan, J. Gutiérrez : "Time dependent solutions for a self gravitating star system". Presented to publication.

17 J.W. Connor and J.B. Taylor : Nuclear Fusion, 17, 1047 (1977)

18 J.W. Connor and J.B. Taylor : Private communication

19 E.D. Courant and H.S. Snyder : Annals of Physics 3, 1 (1958)

20 H.R. Lewis and W.B. Riesenfeld : Journal of Mathematical Physics 10, 1458 (1968)

21 J.R. Burgan, J. Gutiérrez, A. Munier, E. Fijalkow and M.R. Feix : "Group transformations for phase space fluids". Note technique CRPE/47, page 9 (C.R.P.E./P.C.E. Orléans 1977)

22 J.R. Burgan : "Sur des groupes de transformation en physique mathématique" P.H.D. Thesis. Université d'Orléans (1978)

23 J.R. Burgan, M.R. Feix, E. Fijalkow, J. Gutiérrez, A. Munier : "Utilisation des groupes de transformation pour la résolution des équations aux dérivés partielles". Proceedings de la première rencontre interdisciplinaire sur les problèmes inverses. Springer Verlag (1978).

24 G.L. Camm : Self gravitating star systems. Mont. Not. Roy. Soc., 110, 306 (1949) ; 112, 155, (1951).

# NONLINEAR KINETIC EQUATION IN PLASMA PHYSICS LEADING TO SOLITON STRUCTURES

Carlos Montes
Laboratoire de Physique de la Matière
Condensée, Parc Valrose, 06034 Nice
and
Observatoire de Nice, 06007 Nice

Induced Compton scattering of photons – and nonlinear Landau damping of plasmons – by plasma particles are governed by the same type of nonlinear integrodifferential kinetic equation. Numerical time evolution of a large class of initial spectra leads to soliton like behaviour. Only a finite number of exact invariants of motion exist and collisions between solitons are performed in order to test their structural stability.

## I. INTRODUCTION

Up to now, only completely integrable models can be strictly analytically investigated. Only solitary waves satisfying completely integrable equations (which are solved, e.g., by the inverse scattering method) were called "solitons". However, there is a large class of nonintegrable physical systems studied by computer experiments which lead to "soliton like" behaviour. Following Makhankov [1] we may generalize the definition of a soliton as a solitary wave having some "safety factor" which ensures a weak change in interaction or collision with others.

The purpose, here, is to show the time evolution of a large class of initial boson spectra governed by a nonlinear integrodifferential kinetic equation. The numerical computation leads to soliton like behaviour and approximate analytical forms – in the presence or in the absence of a constant noise spectrum – are exhibited. The structural stability is tested by soliton collisions. Paired collisions between boson solitons can take place at a finite range in the same way that particles having a potential. The existence of two exact invariants of motion forces the two outgoing solitons to be equivalent to the incoming ones after exchanging their identities. Multiple soliton collisions, however, violate this ; e.g. ternary collisions show the rise of a much smaller fourth "soliton".

## II. NONLINEAR KINETIC EQUATION

One dimensional induced scattering of bosons (photons or plasmons) by a gas of non relativistic particles is governed by a nonlinear integrodifferential kinetic equation of the form [2-6]

$$\frac{\partial N(x,t)}{\partial t} = N(x,t)\int_0^{\infty} dx'\, W(x,x')\, N(x',t) \tag{1}$$

where $N(x,t)$ is the high level boson occupation number (the spectral maximum satisfying $N_{max} \gg 1$) and where

$$W(x,x') = -W(x',x) \tag{2}$$

is the antisymmetric transition probability kernel. In what follows we shall take

$$W(x,x') = \frac{d}{dx} G(x-x') \tag{3}$$

where $G(x-x')$ is a function tending rapidly to zero for $|x-x'| \to \infty$ with a width $\Delta$ . Then, for $x \gg \Delta$ , the lower bound of the integral in (1) can be replaced by $-\infty$ without appreciable error. This avoids the problem of boson condensation near $x = 0$ . Equation (1) has two exact invariants of motion, namely

$$I_1 = \int_{-\infty}^{\infty} N(x,t)\, dx \tag{4}$$

$$I_2 = \int_{-\infty}^{\infty} \log N(x,t)\, dx \, , \tag{5}$$

and the following expression related to invariant (5)

$$\frac{dJ}{dt} \equiv \frac{d}{dt}\int_{-\infty}^{\infty} x \log N(x,t)\, dx = -\langle G \rangle I_1 \, , \tag{6}$$

where

$$\langle G \rangle = \int_{-\infty}^{\infty} d\xi\, G(\xi) \tag{7}$$

$I_1$ and $I_2$ seem to be the only exact invariants of Eq. (1). Other exact invariants have not been found.

For the case of induced Compton scattering of counterstreaming photons by Maxwellian electrons, $W(x,x')$ is of the form [4,5]

$$W(x,x') = \frac{W_0}{\sqrt{\pi}\,\Delta} \frac{\partial}{\partial x} \exp\left[-\left(\frac{x-x'}{\Delta}\right)^2\right] \tag{8}$$

where $x = \nu/\nu_0$ is the reduced frequency, $\Delta = 2(v_{th}/c)$ is the Doppler width and $W_0 = (n_e \sigma_T hc/m_e c^2)\,\nu_0$ is the Thomson transition probability strength. For the case of nonlinear Landau damping of plasmons by Maxwellian ions, the kernel (8) may be a good approximation [3,7,8] by taking $x = k/k_{De}$, as the reduced wave number ($k_{De}$ being the Debye wave number), $\Delta^2 = m/M$ as the electron/ion mass ratio, and

$$W_0 = (m/M)\left[\hbar\,\omega_p\,k_{De}^3/(2\pi)^3 n_e k T_e\right]\left[T_e T_i/(T_e+T_i)^2\right]\omega_p$$

standing for the normalization of the energy density per dk of the fluctuating electric field by the plasma kinetic energy density ($\omega_p$ being the plasma frequency, $n_e$ the electron density and $T_e$, $T_i$ the electron and ion temperatures).

The parameters $W_0$ and $\Delta$ in kernel (7) are useful for the physical interpretation of the scattering problem, however, Eq. (1) may be transformed into an universal form without any parameter. Indeed, defining

$$\xi = x/\Delta \tag{9}$$

$$\tau = W_0\,t/\Delta \tag{10}$$

and taking into account (8), we obtain

$$\frac{\partial N(\xi,\tau)}{\partial\tau} = \frac{N(\xi,\tau)}{\sqrt{\pi}}\int_{-\infty}^{\infty} d\xi' \left\{\frac{\partial}{\partial\xi}\exp\left[-(\xi-\xi')^2\right]\right\} N(\xi',\tau) \tag{11}$$

## III. BOSON SOLITONS

It has been shown [4,5], by a numerical treatment of Eq.(1) with kernel (8), with and without a source term, that in the presence of a constant noise spectrum $N_0$, whichever nonsingular initial condition $N(\xi,0)$ transforms with time into solitary waves moving downward on the frequency and/or wavenumber axis at constant speed $\sigma$ and constant amplitude $N_m$. A narrow highly populated ($N_{max}/N_0 \gg 1$) initial spectrum, centered at $\xi = \xi_0$ and having a width $\delta \ll 1$, transforms, with time, into one boson soliton, its logarithmic spectral intensity taking the approximate asymptotic form [4,5]

$$\log\left[N(\xi,\tau)/N_0\right] = \left[\log(N_m/N_0)\right]\exp\left[-(\xi-\xi_0+\sigma\tau)^2\right] \tag{12}$$

The speed $\sigma = |d\xi_m/d\tau|$ of the soliton moving downward on the $\xi$ axis (where $\xi_m$ is the abscissa of the maximum) may be found by putting $\partial/\partial\tau = \sigma\,\partial/\partial\xi$ into Eq. (11) and integrating twice with respect to $\xi$: from $-\infty$ to $\xi$ and from $-\infty$

to $+\infty$ , yielding

$$\sigma = \left|\frac{d\xi_m}{d\tau}\right| = \frac{\int_{-\infty}^{\infty} N(\xi,\tau)\,d\xi}{\int_{-\infty}^{\infty} \log N(\xi,\tau)\,d\xi} = \frac{I_1}{I_2} \tag{13}$$

Using the approximate expression (12), the invariant expressions $I_1$, $I_2$ and $\sigma$ are given by

$$I_1 \simeq \sqrt{\pi}\, N_m / [\log (N_m/N_o)]^{1/2} \tag{14}$$

$$I_2 \simeq \sqrt{\pi}\, \log (N_m/N_o) \tag{15}$$

$$\sigma \simeq N_m / [\log (N_m/N_o)]^{3/2} \tag{16}$$

Let us obtain the speed (16) in a different way which will be later useful in the study of the time evolution in the absence of a constant noise spectrum. For $\log (N_m/N_o) \gg 1$, we expand (12) as follows

$$\begin{aligned} N(\xi,\tau) &= N_o \exp\{\log (N_m/N_o) \exp[-(\xi-\xi_o+\sigma\tau)^2]\} \\ &\simeq N_o \exp\{\log (N_m/N_o) [1-(\xi-\xi_o+\sigma\tau)^2] \end{aligned} \tag{17}$$

The narrowness of the soliton profile — its width is smaller by the factor $[\log(N_m/N_o)]^{-1/2}$ than the Gaussian transition probability width — allow us to introduce the limit expression of (17), namely

$$N(\xi,\tau) \to \frac{\sqrt{\pi}\, N_m}{[\log (N_m/N_o)]^{1/2}}\, \delta(\xi-\xi_o+\sigma\tau) \tag{18}$$

into the integrant of equation (11) and integrating over $\xi'$ we obtain

$$\frac{1}{N_o}\frac{\partial N(\xi,\tau)}{\partial\tau} = \frac{N_m/N_o}{[\log (N_m/N_o)]^{1/2}}\frac{\partial}{\partial\xi} \exp[-(\xi-\xi_o+\sigma\tau)^2] \tag{19}$$

By taking into account the soliton form (12), this may be written as an equation of linear uniform motion in the $\xi$ axis

$$\frac{\partial}{\partial\tau} \log [N(\xi,\tau)/N_o] = \sigma \frac{\partial}{\partial\xi} \log [N(\xi,\tau)/N_o] \tag{20}$$

where the soliton speed $\sigma$ turns to be given by (16).

The nonlinear structure of Eq. (11) is responsible for the noise dependence of the soliton motion. Indeed, if the spectrum vanishes the motion stops. Therefore, there is only uniform motion in the presence of an uniform noise spectrum $N_0 \neq 0$, which uniformly ensures the nonlinear and nonlocal coupling of the soliton spectrum with the noise spectrum.

Narrow spectrum evolution :

Figure 1 shows the numerical computation of the time evolution of an initial narrow Gaussian spectrum :

$$N(\xi,0) = N_m \exp\left[-(\xi-\xi_0)^2/\delta^2\right] + N_0 \tag{21}$$

of width $\delta = 1/50$, maximum amplitude $N_m = 10^8$, in the presence of a constant noise spectrum $N_0 = 1$. The spectrum $N(\xi,\tau)$ shows two main kinetic regimes : i) First, the

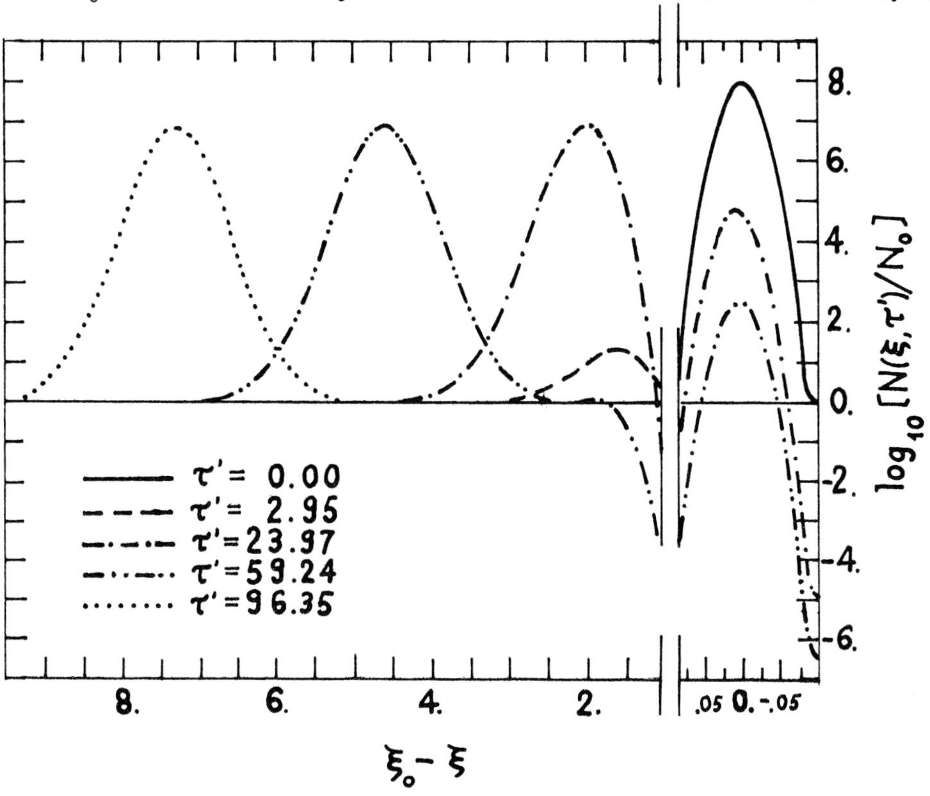


Fig.1 : Time evolution of a narrow Gaussian boson spectrum of width $\delta = 1/50$ and amplitude $N_m = 10^8$ on a noise spectrum $N_0 = 1$ . Logarithmic scale in ordinate ; two different real scales in abscissa. The spectrum becomes with time $\tau' = \tau/\tau_0$ a left satellite and afterwards a soliton of amplitude $N_s = 7.94 \times 10^6$ moving downward on the $\xi$ axis $[\tau_0 = 2^5 \times \sqrt{\pi} \times 10^8]$.

major part of the initial spectrum transforms with time at a finite distance into a satellite on the left wing ; ii) Then, the satellite moves downwards on the $\xi$ axis. It is only in this asymptotic regime that it takes the soliton profile which can be very well approximated in logarithmic scale by a Gaussian, according to the approximate form (12).

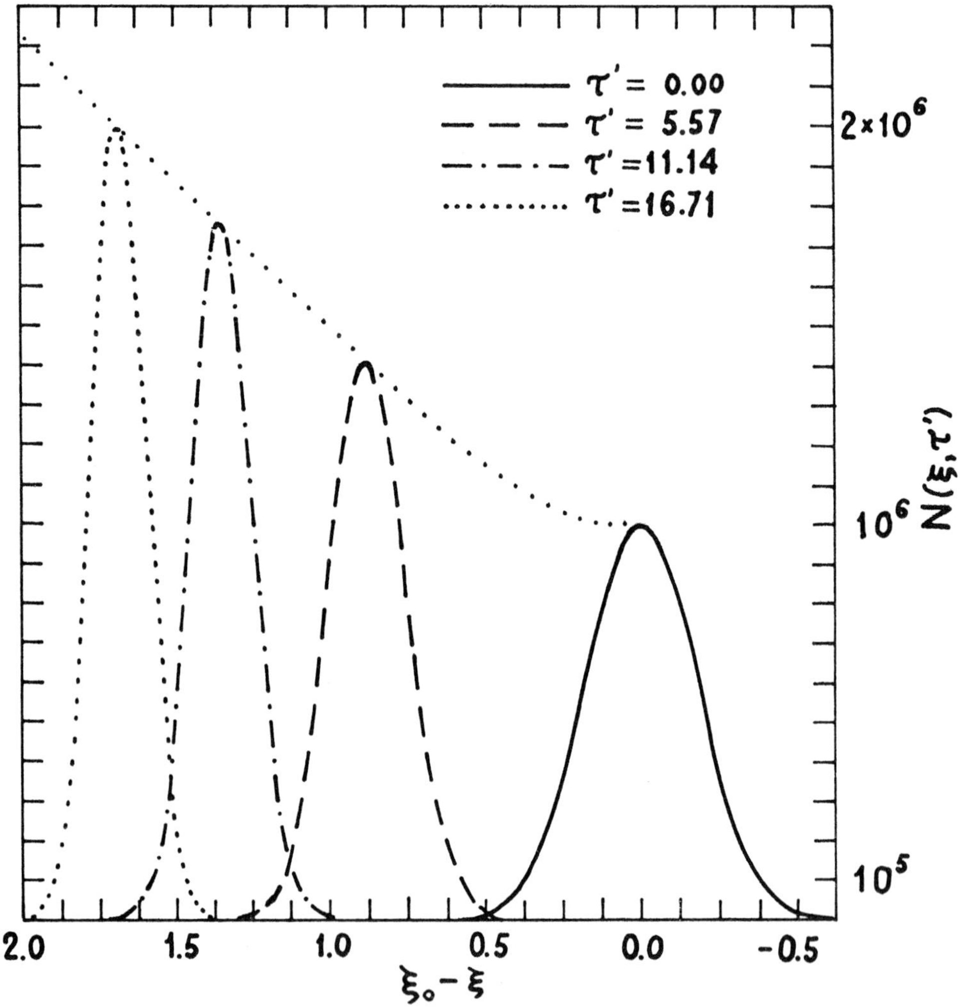


Fig. 2 : Time evolution of a narrow Gaussian spectrum of width $\delta=1/4\sqrt{2}$ and amplitude $N_m=10^6$ in the absence of noise spectrum. Real scales. It moves with time $\tau'=\tau/\tau_o$ downwards on the $\xi$ axis with a decelerated motion, increasing linearly its amplitude (dashed line) and narrowing its width $[\tau_o = 2^5 \times \sqrt{\pi} \times 10^8]$.

In the absence of a noise spectrum $[N_0=0$ in (21)$]$ , the numerical computation of the time evolution of (21) yields a decelerated motion. Figure 2 shows it in real scale. It happens as if the "solitary spectrum" would move on its proper left wing. Its maximum amplitude turns to grow linearly with $\xi - \xi_0$ , its width narrowing and its speed decreasing. We can look for an approximate self-similar form of this solitary spectrum by generalizing the preceding method. We define the function $g(\tau)$ as the primitive of the time dependent speed $\sigma(\tau)$

$$g(\tau) = \int_0^{\tau} \sigma(\tau')\, d\tau' \tag{22}$$

and we introduce the following limit expression for the spectrum

$$\begin{aligned} N(\xi,\tau) &= N_m(\tau) \exp\left\{-\left[\frac{\xi - \xi_0 + g(\tau)}{\Delta(\tau)}\right]^2\right\} \\ &\rightarrow \sqrt{\pi}\,\Delta(\tau)\, N_m(\tau)\, \delta[\xi - \xi_0 + g(\tau)] \end{aligned} \tag{23}$$

into the integrant of Eq. (11). Performing integration with respect to $\xi'$ and with respect to $\tau$ , we obtain

$$\log N(\xi,\tau) = A \int_0^{\tau} \frac{\partial}{\partial \xi} \exp\{-[\xi - \xi_0 + g(\tau)]^2\}\, d\tau + \log N(\xi,0) \tag{24}$$

where A measures the norm, i.e. the constant of motion $I_1$

$$I_1 = \int N(\xi,\tau)\, d\xi = \sqrt{\pi}\,\Delta(\tau)\, N_m(\tau) = \sqrt{\pi}\, A \tag{25}$$

Changing $\partial/\partial\xi$ by $\dot{g}^{-1}\partial/\partial\tau$ in (24), yields

$$\begin{aligned} \log \frac{N(\xi,\tau)}{N(\xi,0)} &= A \int_0^{\tau} \frac{1}{\dot{g}} \frac{\partial}{\partial \tau} \exp\{-[\xi - \xi_0 + g(\tau)]^2\}\, d\tau \\ &= (A/\dot{g}) \exp\{-[\xi - \xi_0 + g(\tau)]^2\}\Big|_0^{\tau} \\ &+ A \int_0^{\tau} \frac{\ddot{g}}{\dot{g}^2} \exp\{-[\xi - \xi_0 + g(\tau)]^2\}\, d\tau \end{aligned} \tag{26}$$

where the dot in g means derivation with respect to $\tau$ . Imposing the conservation of the norm (25) , we are able to obtain a closed differential equation for $g(\tau)$ which solution will give the time dependence of the speed $\sigma(\tau)$ and the self-similar expression for the solitary spectrum. This last, from (26), reads

$$N(\xi,\tau) = N(\xi,0)\exp\left\{\frac{A}{\dot g}\exp\left\{-[\xi-\xi_0+g(\tau)]^2\right\}_0^\tau + A\int_0^\tau \frac{\ddot g}{\dot g^2}\exp\left\{-[\xi-\xi_0+g(\tau)]^2\right\}\right\} \tag{27}$$

Expanding (27) in the vicinity of $\xi-\xi_0=-g(\tau)$ and imposing the conservation of the norm, we obtain

$$[\partial N(\xi,0)/\partial\xi]_{\xi-\xi_0=-g}\,\dot g + N(\xi,0)\frac{\partial}{\partial\tau}\left[\frac{A}{\dot g}\left(1+\frac{\sqrt{\pi}}{2}\frac{\ddot g}{\dot g^2}\right)\right] = 0 \tag{28}$$

which for a Gaussian initial spectrum yields

$$\frac{\partial}{\partial\tau}\left[-\dot g^2 + \frac{A}{\dot g}\left(1+\frac{\sqrt{\pi}}{2}\frac{\ddot g}{\dot g^2}\right)\right] = 0 \tag{29}$$

A first order approximate solution, giving the principal asymptotic dependence of g on $\tau$, may be obtained by neglecting the last term in (29). It yields

$$g = (3A\tau)^{1/3} \tag{30}$$

This simplification is a posteriori justified since the neglected term turns to be $\ddot g/\dot g^2 = -2(3A\tau)^{-1/3}\to 0$ as $\tau\to\infty$. By iteration, we can obtain second order approximations. Numerical computation verifies with good accuracy the derivative of expression (30), namely

$$\sigma(\tau) = \dot g(\tau) = (A/9)^{1/3}\,\tau^{-2/3} \tag{31}$$

Broad spectrum evolution :

Numerical computation of the time evolution of a broad spectrum of the form (21) with $\delta \gg 1$ shows the splitting of the initial profile into secondary ones or solitons[4]. In their motion downward on the $\xi$ axis, they are ordered by their amplitudes, or, that amounts to the same thing, by their speeds, therefore increasing the distances between them. When they are well separated, each soliton approaches the form (12).

## IV. BOSON SOLITON COLLISIONS

In order to test the structural stability of these boson solitons – i.e. of these solitary spectra which move uniformly on a constant noise spectrum – interaction between them has been performed. We recall here the principal results of the computing experiments presented in reference 6 .

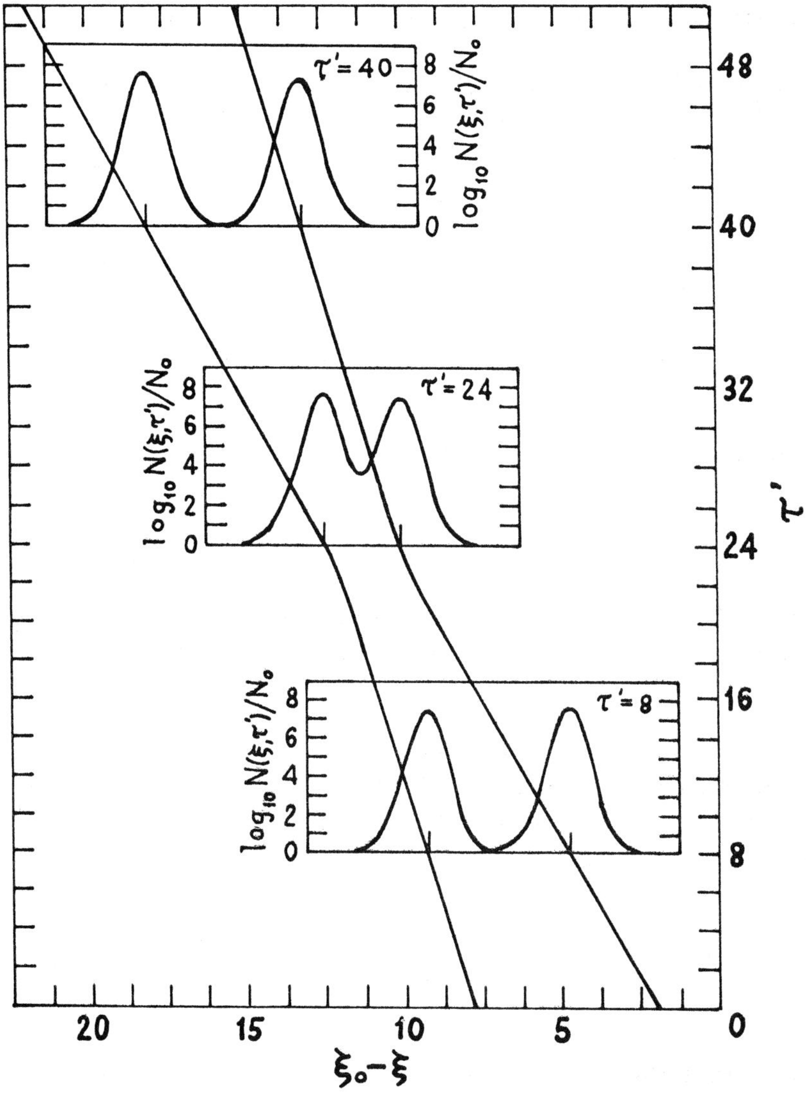


Fig. 3 : Phase diagram $(\xi,\tau')$ of a binary collision between two solitons of amplitudes $N_1 = 4.4678 \times 10^7$ , $N_2 = 2.2766 \times 10^7$ satisfying $N_1/N_2 = 1.9624$.
They completely exchange their identity at a finite interaction range during the encounter. Same time scale as in Figs. 1 and 2.

Binary collisions :

In paired collisions, the solitons may either exchange their identity (strength and speed) at **a** finite interaction range, as particles having a repulsive potential, or may merge by passing through each other when their amplitudes are very different. In all cases there is total transformation : the outgoing solitons seem to be equivalent, within the accuracy of the numerical computation, to the incoming ones. They only experiment an appreciable shift in their motion downward on the $\xi$ axis. The stronger soliton with amplitude $N_1$ exhibits a downward phase shift $\xi_1 - \xi_1'$, whereas the smaller one, with amplitude $N_2$, exhibits an upward phase shift $\xi_2' - \xi_2$, both shifts being related, from Eq. (6) and approximate expression (14), by

$$\frac{\xi_2' - \xi_2}{\xi_1 - \xi_1'} = \frac{\log N_1}{\log N_2} \tag{32}$$

Thus, the existence of the two invariants of motion $I_1$ and $I_2$ forces the outgoing solitons to be equivalent to the incoming ones , if the number of solitons is conserved. Indeed, let $(N_1, N_2)$ and $(N_1', N_2')$ be the amplitudes before and after the interaction of two colliding solitons satisfying $N_1 > N_2 \gg N_o = 1$ . Far enough before and after the collision, we have from (14) and (15)

$$I_1 \simeq \sum_{i=1}^{n} \frac{N_i}{(\log N_i)^{1/2}} = \sum_{i=1}^{n} \frac{N_i'}{(\log N_i')^{1/2}} \tag{33}$$

$$I_2 \simeq \sum_{i=1}^{n} \log N_i = \sum_{i=1}^{n} \log N_i' \tag{34}$$

If n = 2, as $N_1 \gg 1$ and $N_2 \gg 1$, Eqs. (33) and (34) may only be satisfied if $N_1' = N_1$ and $N_2' = N_2$ or $N_1' = N_2$ and $N_2' = N_1$ . This is numerically verified, as we can see on the phase diagram of Fig. 3.

Ternary collisions :

If n=3, the system of Eqs. (33) and (34) does not uniquely determine $N_i'$ (i=1,2,3) by simple permutations. The existence of only two exact invariants of motion cannot constrain the complete permutation of the respective soliton parameters (amplitude and speed) in a ternary encounter, even if the number of solitons is conserved. Numerical computation of the time evolution of three different solitons satisfying Eq. (11) has been performed. By adjusting the impact parameters we obtain the ternary collision represented on Fig. 4 .

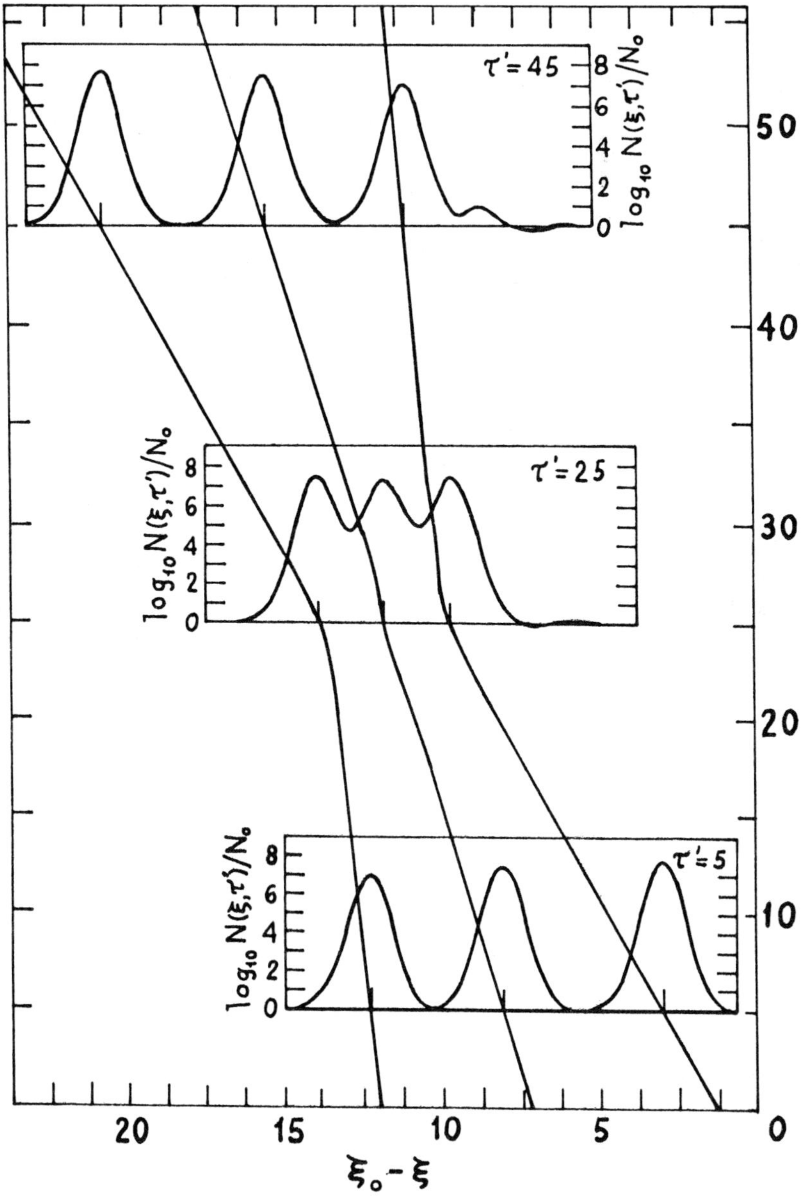


Fig. 4 : Phase diagram of a ternary collision between three solitons of amplitudes $N_1 = 4.4678 \times 10^7$ , $N_2 = 2.2766 \times 10^7$ and $N_3 = 7.4541 \times 10^6$ . There is no total exchange (or permutation) in the encounter. Small deviations from the initial amplitudes and velocities are measurable and what is more important a smaller fourth soliton rises, therefore violating the conservation of the number of solitons. Same time scale as in Figs. 1, 2 and 3 .

We can observe a small deviation of the scattering trajectories with respect to the complete exchange of the incident ones and what is more important the generation of a small fourth solitary spectrum. This spectrum is great enough to become with time (too long for the numerical experiment) a soliton, therefore violating the conservation of the number of such solitons in the interaction.

ACKNOWLEDGEMENTS

I express my gratitude to Professor A.F.Rañada who gives me the possibility to present these results to the assembly. I thank Dr. J. Peyraud for his interest in the problem.

REFERENCES

1. V.G. Makhankov, Physics Rep. 35, 1 (1978).
2. Ya B. Zel'dovich, E.V. Levich and R.A. Syunyaev, Zh. Eksp. Teor. Fiz. 62, 1392 (1972) [Sov. Phys.JETP 35, 733 (1972)].
3. V.E. Zakharov, S.L. Musher and A.M. Rubenchik, Zh. Eksp. Teor. Fiz.69, 155 (1975) [Sov. Phys. JETP 42, 80 (1976)].
4. C. Montes, Plasma Physics : Nonlinear Theory and Experiments (Plenum, New York, p. 222, 1977).
5. C. Montes, Astrophys. J. 216, 329 (1977).
6. C. Montes, J. Peyraud and M. Hénon, Phys. Fluids to be published.
7. R.Z. Sagdeev and A.A. Galeev, Nonlinear Plasma Theory (Benjamin, New York, p. 92, 1969).
8. C. Montes, J. Plasma Phys. 11, 141 (1974).

## Selected Issue[illegible]

# Lecture N[illegible]



Vol. 561: Function Theoretic Methods for Partial Diffe[illegible] Darmstadt 1976. Proceedings. Edited by V. E. Meist[illegible] and W. L. Wendland. XVIII, 520 pages. 1976.

Vol. 564: Ordinary and Partial Differential Equations, Dundee 1976. Proceedings. Edited by W. N. Everitt and B. D. Sleeman. XVIII, 551 pages. 1976.

Vol. 565: Turbulence and Navier Stokes Equations. Proceedings 1975. Edited by R. Temam. IX, 194 pages. 1976.

Vol. 566: Empirical Distributions and Processes. Oberwolfach 1976. Proceedings. Edited by P. Gaenssler and P. Révész. VII, 146 pages. 1976.

Vol. 570: Differential Geometrical Methods in Mathematical Physics, Bonn 1975. Proceedings. Edited by K. Bleuler and A. Reetz. VIII, 576 pages. 1977.

Vol. 572: Sparse Matrix Techniques, Copenhagen 1976. Edited by V. A. Barker. V, 184 pages. 1977.

Vol. 579: Combinatoire et Représentation du Groupe Symétrique, Strasbourg 1976. Proceedings 1976. Edité par D. Foata. IV, 339 pages. 1977.

Vol. 587: Non-Commutative Harmonic Analysis. Proceedings 1976. Edited by J. Carmona and M. Vergne. IV, 240 pages. 1977.

Vol. 592: D. Voigt, Induzierte Darstellungen in der Theorie der endlichen, algebraischen Gruppen. V, 413 Seiten. 1977.

Vol. 594: Singular Perturbations and Boundary Layer Theory, Lyon 1976. Edited by C. M. Brauner, B. Gay, and J. Mathieu. VIII, 539 pages. 1977.

Vol. 596: K. Deimling, Ordinary Differential Equations in Banach Spaces. VI, 137 pages. 1977.

Vol. 605: Sario et al., Classification Theory of Riemannian Manifolds. XX, 498 pages. 1977.

Vol. 606: Mathematical Aspects of Finite Element Methods. Proceedings 1975. Edited by I. Galligani and E. Magenes. VI, 362 pages. 1977.

Vol. 607: M. Métivier, Reelle und Vektorwertige Quasimartingale und die Theorie der Stochastischen Integration. X, 310 Seiten. 1977.

Vol. 615: Turbulence Seminar, Proceedings 1976/77. Edited by P. Bernard and T. Ratiu. VI, 155 pages. 1977.

Vol. 618: I. I. Hirschman, Jr. and D. E. Hughes, Extreme Eigen Values of Toeplitz Operators. VI, 145 pages. 1977.

Vol. 623: I. Erdelyi and R. Lange, Spectral Decompositions on Banach Spaces. VIII, 122 pages. 1977.

Vol. 628: H. J. Baues, Obstruction Theory on the Homotopy Classification of Maps. XII, 387 pages. 1977.

Vol. 629: W. A. Coppel, Dichotomies in Stability Theory. VI, 98 pages. 1978.

Vol. 630: Numerical Analysis, Proceedings, Biennial Conference, Dundee 1977. Edited by G. A. Watson. XII, 199 pages. 1978.

Vol. 636: Journées de Statistique des Processus Stochastiques, Grenoble 1977, Proceedings. Edité par Didier Dacunha-Castelle et Bernard Van Cutsem. VII, 202 pages. 1978.

Vol. 638: P. Shanahan, The Atiyah-Singer Index Theorem, An Introduction. V, 224 pages. 1978.

[illegible] Nonlinear Partial Differential Equations and Applications, Proceedings, Indiana 1976–1977. Edited by J. M. Chadam. VI, 206 pages. 1978.

Vol. 650: C*-Algebras and Applications to Physics. Proceedings 1977. Edited by R. V. Kadison. V, 192 pages. 1978.

Vol. 656: Probability Theory on Vector Spaces. Proceedings, 1977. Edited by A. Weron. VIII, 274 pages. 1978.

Vol. 662: Akin, The Metric Theory of Banach Manifolds. XIX, 306 pages. 1978.

Vol. 665: Journées d'Analyse Non Linéaire. Proceedings, 1977. Edité par P. Bénilan et J. Robert. VIII, 256 pages. 1978.

Vol. 667: J. Gilewicz, Approximants de Padé. XIV, 511 pages. 1978.

Vol. 668: The Structure of Attractors in Dynamical Systems. Proceedings, 1977. Edited by J. C. Martin, N. G. Markley and W. Perrizo. VI, 264 pages. 1978.

Vol. 675: J. Galambos and S. Kotz, Characterizations of Probability Distributions. VIII, 169 pages. 1978.

Vol. 676: Differential Geometrical Methods in Mathematical Physics II, Proceedings, 1977. Edited by K. Bleuler, H. R. Petry and A. Reetz. VI, 626 pages. 1978.

Vol. 678: D. Dacunha-Castelle, H. Heyer et B. Roynette. Ecole d'Eté de Probabilités de Saint-Flour. VII-1977. Edité par P. L. Hennequin. IX, 379 pages. 1978.

Vol. 679: Numerical Treatment of Differential Equations in Applications, Proceedings, 1977. Edited by R. Ansorge and W. Törnig. IX, 163 pages. 1978.

Vol. 681: Séminaire de Théorie du Potentiel Paris, No. 3, Directeurs: M. Brelot, G. Choquet et J. Deny. Rédacteurs: F. Hirsch et G. Mokobodzki. VII, 294 pages. 1978.

Vol. 682: G. D. James, The Representation Theory of the Symmetric Groups. V, 156 pages. 1978.

Vol. 684: E. E. Rosinger, Distributions and Nonlinear Partial Differential Equations. XI, 146 pages. 1978.

Vol. 690: W. J. J. Rey, Robust Statistical Methods. VI, 128 pages. 1978.

Vol. 691: G. Viennot, Algèbres de Lie Libres et Monoïdes Libres. III, 124 pages. 1978.

Vol. 693: Hilbert Space Operators, Proceedings, 1977. Edited by J. M. Bachar Jr. and D. W. Hadwin. VIII, 184 pages. 1978.

Vol. 696: P. J. Feinsilver, Special Functions, Probability Semigroups, and Hamiltonian Flows. VI, 112 pages. 1978.



Vol. 702: Yuri N. Bibikov, Local Theory of Nonlinear Analytic Ordinary Differential Equations. IX, 147 pages. 1979.

Vol. 704: Computing Methods in Applied Sciences and Engineering, 1977, I. Proceedings, 1977. Edited by R. Glowinski and J. L. Lions. VI, 391 pages. 1979.

Vol. 710: Séminaire Bourbaki vol. 1977/78, Exposés 507–524. IV, 328 pages. 1979.

Vol. 711: Asymptotic Analysis. Edited by F. Verhulst. V, 240 pages. 1979.